# Do Your Ears Pop in Space?

**Also by R. Mike Mullane**

*Red Sky, A Novel of Love, Space & War*

*Liftoff! An Astronaut's Dream*

# Do Your Ears Pop in Space? and 500 Other Surprising Questions about Space Travel

R. Mike Mullane, Astronaut

**John Wiley & Sons, Inc.**
New York • Chichester • Brisbane • Toronto • Singapore • Weinheim

Copyright © 1997 by R. Mike Mullane

**Library of Congress Cataloging-in-Publication Data**
Mullane, R. Mike.
  Do your ears pop in space? and 500 other surprising questions
  about space travel / R. Mike Mullane.
      p.   cm.
  Includes index
  ISBN 0-471-15404-0 (pbk. : alk. paper)
      1. Astronautics—Miscellanea.    I. Title.
TL793.M7893      1997
629.4—dc 20                                          96-28544
                                                        CIP

Printed in the United States of America

20 19 18 17 16 15 14 13

This book is dedicated to the heroes and heroines of America's space program—the NASA team.

# Acknowledgments

Believe it or not, astronauts don't know everything about space, and I depended upon many other people to complete this work. A sincere thanks goes to the following individuals who gave freely of their time to answer questions or review my manuscript: Steve Nesbitt, Pam Alloway, Doug Ward, and Norma Rhoads of the NASA Johnson Space Center (JSC) Public Affairs Office; other NASA JSC employees Dan Adamo, Karen Ross, Duane Ross, Karen Edelstein, David Shaw, Bill Atwell, Bob Williams, Sharon Jones, Olan Bertrand, and Bill Bates; Astronauts Sid Gutierrez, Pierre Thuot, Jerry Ross, Robert (Hoot) Gibson, Jim Bagian, Dick Truly, Steve Hawley, Ken Bowersox, and Rhea Seddon; Dennis Stocker of NASA's Lewis Research Center in Cleveland, who was particularly helpful in reviewing my comments on weightlessness; Dr. Joe Boyce, a former NASA flight surgeon, and Dr. Denise Baisden, a current flight surgeon, who were my medical sources; Tammy West, the source of some of my astronaut facts (as an astronaut office secretary, she probably knows more about astronauts than anybody alive); Colin Sikorski of Hitachi Semiconductor (America) Inc., who did some Einsteinium relativity calculations for me; Kay Hemme of Delta Airlines, who deserves a huge thanks for making it possible for me to travel so extensively to document Americans' curiosity about space. Another big thank you goes to my children — Patrick, Amy, and Laura — who have always been enthusiastic audiences for my literary efforts. Finally, a heartfelt thanks goes to my wife Donna. She's been my best friend and confidant for 29 years. Thanks again, everybody!

# Contents

# Preface

In the early morning hours of February 28, 1990, I was strapped into the cockpit of the space shuttle *Atlantis* awaiting my third and final mission into space. My body was in agony. We were at T minus 9 minutes and holding. Bad weather had already postponed the launch for an hour beyond the intended lift-off time, and we had been in the cockpit for nearly 3 hours. The steel chair, lumpy parachute, and stiff pressure suit were all conspiring to torture my back. I arched my body to relieve some of the worst pressure points. The 83 pounds of equipment that enveloped me made the movement a struggle, but the momentary restoration of circulation was heaven-sent. But the relief was brief. As my muscles fatigued and my body sagged back to the seat, I could feel a puddle of cold urine rising around my crotch. It was being squeezed from my urine collection device (the astronaut name for a diaper). "Now I know why babies cry when their diapers are wet," I thought. The feeling was disgusting.

My discomfort was forgotten as the Launch Director radioed that the weather was improving. We were coming out of the hold. T minus 9 minutes and counting. Once again the digits began to evaporate from the countdown clock and gut-level fear seized my soul.

At T minus 5 minutes, the pilot started the shuttle's hydraulic pumps. I felt a slight vibration in the cockpit as the fluid began to course through her veins.

At T minus 2 minutes, I closed my helmet visor and switched

on my emergency oxygen. I could now *hear* my fear in the hissing sound of my accelerated breathing.

The last call came from the Launch Director: "Good luck and godspeed, *Atlantis*." T minus 30 seconds. Except for the soft, whooshing background noise of the cabin fans, the cockpit was deathly quiet.

T minus 10 seconds. My heart choked me. I tried to swallow, but I had no saliva.

T minus 9. . .8. . .7. . . .

I think it's a safe bet, you're curious as to what happened next. What does a shuttle launch feel like? Sound like? How many g's do you experience? What does the earth look like from space? Why are you weightless in space? How do you go to the bathroom in a shuttle? Have I seen any UFOs? These are just a handful of the thousands of questions I've been asked by audiences around the country. As I said, I would be surprised if you didn't share a similar curiosity. Until now, however, there was no single, comprehensive and authoritarian source from which to find answers to such questions. *Do Your Ears Pop in Space? And 500 Other Surprising Questions about Space Travel* fills that void. Turn to Chapter 2 to have your questions answered about the sensations of a shuttle launch, to Chapter 4 to learn what it's like to sleep in space, to Chapter 5 to see if your ears really *do* pop in space, or to Chapter 8 to find out how much money astronauts are paid.

This is a *unique* work. While other astronauts have written similar-themed books, none have had the opportunity to compile such an extensive list of laypersons' questions. My opportunity to do so has come from the fact that my post-astronaut business has been as a full-time professional speaker. I know of no other astronaut with a similar business. Certainly, other astronauts give speeches, but they do so infrequently and as an adjunct to their primary businesses. For me, professional speaking has been my entire post-astronaut life, and in the past 6 years I have compiled over a million frequent flyer miles while traveling to address hundreds of thousands of people of all ages and from all segments of our society. My audiences have included the employees of Fortune 500 companies, students (from kindergarten through college), associations, societies, fraternal organizations, chambers of commerce, retirement organizations, future farmers, scouts, and countless others. Some of these audiences have exceeded 3,000 people and others have totaled less than 30. But, regardless of the affiliation, age, gender, or size of the audience, they have all shared one characteristic: They have wanted to know why and how and what if. They have peppered me with thousands upon thousands of questions. It's been this unique exposure that has made me an expert on the public's curiosity about space and prompted me to write this book.

This is a down-to-earth book in which I have addressed the 500 questions I've been most frequently asked about space travel and the astronaut experience. The answers are given in an easy-to-read, entertaining, conversational, and fun manner. You don't need to be a rocket scientist to be entertained and educated by this book. Anybody who has had even a passing interest in the space program will be intrigued by the answers. Some of these will correct popular media misinformation, for example, that astronauts are weightless because there's no gravity (incorrect) or the only human-made object astronauts can see from space with the naked eye is the Great Wall of China (also wrong) or that the military has spy satellites parked over Iran (absolutely wrong). Teachers and students of all ages will find the book an invaluable reference source. When kids want to know what it is like to bleed in space or what do the planets look like from a shuttle or why the *Challenger* blew up, this book will have a ready, easy-to-understand answer.

While I'm very confident that this book answers the most popular questions about space and astronauts, I'm certain there are others I've yet to hear. Please E-mail any additional questions to mikmullane@aol.com.

I will not answer the questions on the Internet, but I will consider your questions for possible inclusion in the next revision of this work. Also, if you care to order an autographed copy of any of my books, please E-mail your name and address to the same Internet address or write to me care of the publisher, and I'll forward an order form to you.

One final comment about this work. To avoid having to use the awkward combination of he/she and him/her to gender neutralize the book, I've merely mixed my gender references. In some answers I refer to the subject as female and in others I use male references. In reality, every astronaut position has male and female representation.

Enjoy.

R. Mike Mullane
NASA Astronaut (Ret.)

# About the Author

Astronaut Mike Mullane was selected in the first group of space shuttle astronauts in 1978 and spent 12 years with NASA. He is a veteran of three space shuttle missions and has logged 356 hours in space aboard the shuttles *Discovery* (STS-41D) and *Atlantis* (STS-27, STS-36). A 1967 graduate of West Point, Colonel Mullane also holds a Master's of Science Degree in Aeronautical Engineering from the Air Force Institute of Technology. In his Air Force career he compiled 150 combat missions in Vietnam and over 3000 hours of flying time in fighter aircraft. Since his retirement from NASA and the Air Force in 1990, he has written a hardcover adult novel and an award-winning children's book and is self-employed as a professional speaker. He resides in Albuquerque, New Mexico.

# CHAPTER 1

# Space
# Physics

## Is there gravity where a shuttle orbits?

Absolutely! In fact, it's gravity that keeps the shuttle in orbit. Without gravity, it would fly into deep space and never return to Earth.

Gravity does get weaker as you travel away from the earth (or any object with mass, for that matter). The exact amount of loss is a function of this simple relationship:

$$\frac{1}{R^2}$$

The earth is about 4,000 statute miles (we'll use statute miles throughout this book because that's how most Americans think of miles) in radius from its core to its surface, where we live. ($R = 1$ at the earth's surface). When standing on the earth's surface, we say we are at "1" gravity. This relationship simply states that if you travel from the center of the earth to a distance two times the earth's radius, or 4,000 miles above the earth (where $R = 2$), the gravity doesn't drop to one half but rather to one fourth or $1/2^2$. So, if you could climb a tower that was 4,000 miles high and could stand on a scale, it would show one fourth of your Earth weight. But shuttle astronauts don't orbit at 4,000 miles altitude. They circle about 200 miles up. This is an R of only 1.05. According to the preceding relationship, this means a shuttle orbits in a gravity field that's about 91% of Earth's surface gravity. Put another way, if there were a 200-mile-high tower and you climbed to its top and weighed yourself, you would see 91% of your Earth weight. Gravity has dropped by only 9%.

## Why are astronauts weightless?

From the answer to the first question, it's obvious astronauts are not weightless because there's no gravity. Astronauts are weightless because they are freely under the influence of gravity. The common term used to describe this condition is *free fall*. Let me explain. Suppose I was standing on a scale in an elevator that's at the top of a skyscraper. I would see my weight on that scale

(160 pounds). Now, suppose someone cuts the elevator cable. What would I see on the scale in my free fall down the elevator shaft? Before you answer, think about what's happening. The elevator floor is falling. The scale is falling. I'm falling. Everything is falling (accelerating) freely (if you ignore the effects of air friction and other types of friction). I'm no longer *standing* on the scale. I'm falling with it. So, it's going to read zero. I'm weightless. Simply put, any time you are able to move freely in response to gravity—when there is nothing to restrain you from accelerating or decelerating with it—you are weightless. When you think about it, the only reason anything weighs anything on Earth is because the ground gets in the way of what gravity wants to naturally do—pull us to the center of the earth. When you stand on a scale, the weight you see is merely the equal and opposite reactions of the earth getting in the way of a fall.

## Why don't astronauts hit the ground in their free fall?

Let's suppose there's a skyscraper that's 200 miles high (a typical shuttle altitude). Let's further suppose I could shrink you down to the size of an ant and put you inside a windowless box. Now, I take that box to the top of the building and drop it over the side. Your fall would be identical to my hypothetical elevator ride. As soon as I released

*Trajectories vary as the speed of objects is increased.*

the box, you would be weightless, because the floor would no longer be in the way of what gravity wants to do—pull you down. In other words, the bottom of the box would be falling out from under your feet as rapidly as your body is falling. Suppose, now, instead of dropping the box, I threw it horizontally from the edge of the building. What would that ride feel like?

Momentarily you would be pinned to the wall as my arm accelerated you; that is, you would feel g-forces, just like an astronaut feels when the engines are thrusting. But as soon as my pitching force ended, you would again be freely under the influence of gravity and would once again be weightless. Now, however, your fall wouldn't be straight down. My throw has given you a constant horizontal velocity away from the building (remember, we are assuming no sources of friction), while, at the same time, gravity is pulling you and the box downward. It's important to understand that the horizontal velocity away from the building is not affected by the downward acceleration of gravity. At any instant you are continuing to move horizontally away from the building at the exact speed you had when the box left my hand. It's just that you are also continually gaining downward speed from the pull of gravity. This combination of horizontal speed and downward acceleration results in a curved trajectory. Would you have any idea that this time, you would be falling in a curve? No. The box is windowless. All you would know is that you are weightless. Now, if I successively increased the power of my throw, I could send you into a weightless flight that would take you further and further from the building, and each time, you would have no idea what trajectory you were tracing. Suppose, however, I could throw the box at a speed of 17,300 mph. As soon as the impulse of that throw ended you would again be in weightless flight, but now the curve is flattened so much that it traces the curvature of the earth. In other words, I've thrown you into orbit. You would still be freely under the influence of gravity, just as you were when I dropped the box straight down and after each of my lower velocity pitches, but now you won't hit the ground. You're in orbit.

### Is orbit speed always 17,300 mph?

No. Orbit velocity depends on orbit altitude and the mass of the planet being orbited (how much stuff it's made of). To orbit something at a low altitude, you must go faster because gravity is stronger. Similarly, the more massive an object is, the stronger its gravity and the faster you have to go to orbit it. An orbit just above Jupiter's clouds would require a speed of 94,300 mph, while orbit velocity just above the surface of our moon is only 3,760 mph. On a very small asteroid, orbital velocity would be even less. In fact, you might be able to find a tiny asteroid that has gravity so weak that its orbit velocity is only 50 mph. On such an object, it would be possible to play catch with yourself. You could throw a ball, watch it disappear over the horizon, and then turn around and catch it as it circled in orbit.

### What is a Vomit Comet?

Remember, the only way you can be weightless is to be freely under the influence of gravity, and you can do that in a falling airplane. To prepare astronauts and shuttle experiments for weightlessness, the National Aeronautics and Space Administration (NASA) flies an old Boeing passenger jet on a roller coaster trajectory. At the top of each hill the pilot pushes the nose of the airplane downward. This effectively drops the floor from underneath the passengers and allows them to be freely under gravity's influence. When this happens everything inside the jet is momentarily weightless. Unfortunately, as the nose gets too low, the pilot has to pull out of the dive, and everything in the aircraft is subsequently pinned to the floor with a force of about 1.8 g's (nearly twice the pull of normal gravity). Then, the pilot climbs back up and repeats the maneuver. After about an hour of this roller coaster ride, you'll understand why the plane has been nicknamed the Vomit Comet. It is a nauseating ride, and many people get air-sick.

### Besides the Vomit Comet, is there another way of duplicating weightlessness on Earth?

Yes. Because weightlessness is a free fall, dropping something down a deep hole would also make it temporarily weightless. Believe it or not, NASA has a 430-foot-deep hole in the ground at the Lewis Research Center in Cleveland where things can be

**Astronauts floating inside the Vomit Comet.**

dropped to measure their response to weightlessness. To remove the effects of air friction, which ruins perfect weightlessness, air is pumped out of the hole to create a vacuum. Experiments are then literally dropped down this hole. High-speed data collection equipment enables engineers to evaluate experiments for about 5 seconds of true weightlessness before a bed of styrofoam cushions the impact. Fortunately, NASA doesn't drop astronauts down this hole for their training.

NASA can double the weightless test time in this hole by launching the experiment upward with compressed air from the bottom of the hole instead of dropping it from the top. While it's not intuitively obvious that something moving *upward* could be as weightless as something falling *downward*, it is a fact. Visualize this: With a scale attached to my feet, I'm launched upward from the bottom of the hole by compressed air. After the launch impulse ends, what does the scale show? Think about it. The scale is being slowed down by the pull of gravity at exactly the same rate my body is being slowed. Therefore, I'm not standing on it. The scale and I are in a weightless free fall, except, this time, we're moving upward. I would continue to slow down until I reached zero speed at the very top of the hole, then I would begin the fall downward. Throughout the entire flight I would be weightless because I would be freely under the influence of gravity. This is why pioneering astronaut Alan Shepard was weightless—during his first mission. He didn't get into orbit on his first flight. It was a 15-minute "lob" out into the ocean. As soon as his Redstone rocket's engine shut down, however, he was freely under the influence of gravity and was therefore weightless—even though the capsule was still moving away from the earth. Similarly, shuttle astronauts first experience weightlessness when their main engines shut down—even though they are still flying upward toward apogee (the high point of the orbit). Remember, weightlessness has nothing to do with the direction you're moving in. It occurs whenever gravity is freely able to do what gravity wants to do—pull you. It doesn't matter if that pull is speeding you up as you fall toward Earth or is slowing you down as you are moving away from Earth. In both cases, as long as nothing is getting in the way of that invisible pull, you'll be weightless.

## What is microgravity?

True weightlessness can never be achieved aboard the shuttle. There is always a very small force—a deceleration—produced by atmospheric drag. This force is measured in *millionths* of an Earth g and is referred to as microgravity. Because it's so small, astronauts can't feel this force.

## Why do astronauts go underwater to train for spacewalks?

Many people believe we do this because we're weightless under water. This is not correct. Are people in submarines weightless? Of course not. Think of underwater astronauts as being in very small submarines having the shape of a space suit. Like the Navy seaman in a real submarine, they're not weightless. So, why do we practice space walks underwater?

**Astronauts train for Hubble spacewalk underwater.**

To understand the answer to this question, you must first understand what happens when a spacewalking astronaut tries to work on a satellite in orbit. If she is not restrained, her body will move in the *opposite* direction of any force she exerts. In other words, if an astronaut has a wrench and is trying to loosen a bolt and exerts a counter-clockwise force, the bolt will not loosen. Instead, the astronaut's body will turn in the opposite direction—clockwise. Now, if you're *neutrally buoyant* underwater (meaning you're not sinking and not rising, just floating at the same depth), and you tried to loosen the same bolt on an underwater mock-up of the satel-lite, you would get the same reaction. Your body would move *opposite* to any force you exert. It is this effect that makes prac-ticing underwater a good preparation for a spacewalk. Doing the work underwater shows the engineers how tools need to be de-signed, where handholds and foot restraints need to be placed, and it allows the astronaut to practice doing the work inside a bulky suit.

### Can you weigh yourself in weightlessness?

In weightlessness it's impossible to weigh anything, yet in some life-science experiments, changes in body weight are im-portant research parameters. How can weight be measured? NASA makes use of Newton's second law to record changes in weight:

$$F = ma$$

This equation states that *force* (the weight you see on the bath-room scale) is equal to your *mass* times the *acceleration* (gravity). So, in space, if you could determine your mass, it would be easy to calculate how your Earth weight is varying. A special NASA device does just that. When it's necessary to see how his Earth weight is changing, an astronaut straps his body to a chair that is accelerated along a short rail by a known spring force ($F$ in the equation). During the motion, acceleration is measured by special electronics. Because $F$ is known and $a$ is measured, it's easy to then calculate the mass of the person riding the chair. Using that mass figure and Earth's gravity for acceleration, the same equation ($F = ma$) can be used to calculate a person's Earth weight.

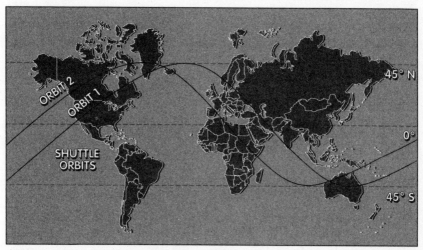

**If a shuttle is orbiting in a circle, why does the Mission Control map show wavy lines?**

The wavy lines you see on Mission Control's map are ground tracks. They show the path of the space shuttle over the ground and look like this:

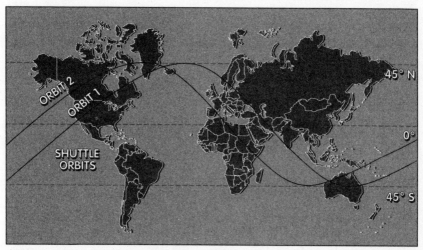

*A shuttle orbit is approximately a circle, but on a flat map it looks like a wavy line.*

The lines appear wavy because of the effect of trying to project an inclined *great circle* (a shuttle orbit) onto a flat map.

Try this experiment. You'll need an Earth map, an Earth globe, and some rubber bands. The map and globe need to have latitude and longitude lines. Tie the rubber bands into a long chain and then tie each end of the chain together. Now you have a chain loop of rubber bands. Stretch that loop over the globe at an angle of about 45 degrees to the equator. (Make certain the loop traces a great circle, that is, its imaginary center is the center of the globe.) We'll pretend this is a shuttle orbit. Now, look at where your rubber band loop crosses the equator and record the latitude and longitude (the latitude will be zero at the equator). Go about 2 inches along your orbit (the rubber band) and, again, write down the latitude and longitude of that point. Keep doing this until you've recorded the latitude and longitude at about 2-inch segments all the way around your orbit. Now, plot each of those points on your Earth map and connect the dots. What do you have? You have a wavy line just like that completed by Mission Control. Your rubber band is a circle, just like a shuttle orbit, but when you plot it out on a flat map, it looks like a wavy line.

### Does a shuttle orbit continually trace the same path across the earth?

No. The shuttle orbit is fixed in space and never changes, but the Earth spins underneath it. This means the orbit passes over different spots on the earth. Let's say you are a shuttle astronaut and look out the window as you cross the equator and see Lake Victoria. (Look at a map. Lake Victoria is in the east African country of Kenya. It's a big lake and easily seen from orbit.) Ninety minutes later you are again crossing the equator over Africa, but now Lake Victoria is about 1,553 miles to the east. Some simple math explains this change in ground track. (I've ignored the minor effects of earth's precession.)

In one day (in 24 hours), the earth makes one complete 360-degree turn to the east. That means it's spinning 15 degrees per hour:

$$360 \text{ degrees}/24 \text{ hours} = 15 \text{ degrees per hour}$$

At the equator, each degree is equal to 60 nautical miles (about 69 statute miles), so, every hour, something on the equator, like Lake Victoria, is going to move about 1,035 miles to the east.

$$15 \text{ degrees} \times 69 \text{ miles per degree} = 1,035 \text{ miles (approximately)}$$

But your orbit wasn't 1 hour. It was 1.5 hours (90 minutes). So, when you next cross the equator, Lake Victoria is going to have moved 22.5 degrees, or about 1,553 miles, eastward.

$$22.5 \text{ degrees} \times 69 \text{ miles per degree} = 1,553 \text{ miles}$$
$$(\text{approximately})$$

In other words, every time astronauts cross the equator, it appears they have moved about 1,553 miles to the *west* when, actually, their orbit hasn't changed at all. It's the earth that has moved. At the equator, it has spun about 1,553 miles to the *east*.

### Why are shuttles launched only from Florida?

There are two reasons. One is safety. You don't want to endanger people on the ground by launching a rocket over their heads, so we launch them over water. Also, for the shuttle, the booster rockets need to parachute into the ocean to be picked up and reused. An impact on land would destroy them. But there is also a space physics reason for launching from Florida. When you do so, you get a free ride from Mother Nature. Remember, the earth is spinning to the east. That spin represents free velocity to a rocket headed eastward into orbit.

To better understand this, let's go back to the equator. There, *everything* is spinning toward the east at 1,035 mph. The trees, people, buildings, the air, the birds, *everything*—including a rocket on a launch pad—is traveling 1,035 mph due eastward. In other words, a rocket on an equatorial launch pad hasn't even lifted off and it already has 1,035 mph toward its orbit speed of 17,300 mph. You would be foolish not to take advantage of this free speed and launch your rocket to the east. That's the space physics reason why NASA launches the shuttle eastward from Florida. To do so reduces the fuel load and increases the payload weight that can be carried.

Having said this, however, the shuttle doesn't lift off with a free 1,035 mph. Remember, that's the earth's spinning speed at the *equator*. The Kennedy Space Center isn't on the equator. It's at approximately 28.5 degrees north latitude, so its spinning speed is less.

To better understand this, take a look at a globe. As you move away from the equator, the circles formed by lines of latitude become smaller. Regardless of latitude, though, it still requires 24 hours to spin the complete circle. This means, the further you move away from the equator, the slower your spinning speed. For example, if you were on the line of latitude that's only a foot southward from the north pole, you would spin a circular distance of about 6 feet in 24 hours, for a speed of about 3 *inches* per hour. The latitude of the Kennedy Space Center measures a circular distance only 88% of the equator's circle, so the Kennedy Space Center's spin speed is only 88% of the equator's speed, or about 911

*The eastward spin velocity of the earth is slower at latitudes away from the equator.*

mph. It doesn't take a rocket scientist to figure out that it would be much better to build a launch site in Florida than in Maine.

By the way, what country has a launch pad best situated to take advantage of Mother Nature's free spin? France. France built its Ariane launch pad in French Guiana, South America (latitude about 5 degrees). The Russians, because of the high latitude of their land mass, have the least favorable launch site—the Baikonur Cosmodrome (east of the Aral Sea, in Kazakhstan) at a latitude of 45.9 degrees north.

### Why does the shuttle roll after lift-off?

This is because sometimes we don't want to launch due east. Even though this is the best direction to maximize the free boost of the earth's spin, such a launch trajectory might not support the mission objective. This statement is best explained by example.

If you were a scientist and had a shuttle experiment designed to study the ozone hole over the Antarctic, which of the orbits shown in the picture on page 13 would you prefer to fly? Obviously you would want the *near-polar* orbit—the orbit that takes you close to the north and south poles. Otherwise your instruments would never even see the ozone hole. For the same rea-

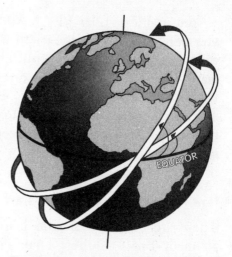

son, if you were in the military and wanted to launch a
spy satellite, you would de-
sire the near-polar orbit. As
the earth spun underneath
you, you would eventually
see most of the earth and
have a view of all potential
enemies. But what if you
were launching a communi-
cation satellite? Most of
these types of satellites are
aimed for orbits over the
equator. Clearly, in that case
you would want the rocket to
fly the orbit closest to the

**The tilt of an orbit toward the equator is its inclination.**

equator. Or what about rendezvousing with the *Mir* space station? Its orbit is tilted 51 degrees to the equator.

It's the *mission* that dictates what orbit a shuttle will fly. The
tilt of that orbit to the equator is called its *inclination*. In turn,
the launch direction is dictated by this inclination. If the mission requires a high-inclination orbit (tilted toward the poles),
the shuttle will launch to the northeast. If the mission requires
a minimum-inclination orbit, the shuttle will fly due east.

It's this need to vary the orbit inclination from mission to
mission that makes it necessary for the shuttle to roll on its tail
after lift-off. The roll aligns the shuttle's pitch axis correctly so
that when it begins to tilt (pitch) over, its nose is aimed in the
right direction for its journey into orbit. Because the shuttle sits
on the launch pad with its tail fin pointed south (and a shuttle
will never fly south into orbit), some type of a roll is always necessary. After lift-off, if the shuttle is destined for a minimum-
inclination orbit, it will roll until the tail is pointed due
eastward and then begin its pitch over to aim the nose eastward.
If it's aiming for a high-inclination orbit, it will roll until the tail
is pointed northeastward (parallel with the east coast of the
United States) and then begin its pitch over.

You might wonder why NASA built the launch pad so the shuttle's tail points to the south when it's awaiting launch. Why didn't they build it so the tail points eastward and thus minimize the need to roll? This arrangement was due to the fact that the shuttle pads weren't originally built for the shuttle. They are converted Saturn V launch pads from the Apollo program. The conversion did not permit a tail-east shuttle launch orientation.

## How many shuttle launch pads are there?

There are only two shuttle launch pads and both are at the Kennedy Space Center. Their designations are pads 39A and 39B. These are more than sufficient to support the shuttle launch schedule of seven or eight missions per year.

## Can the shuttle fly over the north and south poles?

Many scientists—for example, geologists, oceanographers, and meteorologists—as well as military forces want their satellites to be able to see the entire earth's surface. The only orbit that permits such a view is a polar orbit. Unfortunately, the shuttle cannot launch satellites into such orbits. For safety reasons, NASA will not launch the shuttle over land, so a shuttle launching into the highest possible inclination will fly northeast heading just off the coast of the United States and achieve a 62-degree inclination in the process. Higher inclinations are safely achieved by unmanned rockets flying southward over the Pacific Ocean from Vandenberg Air Force Base in California. Originally, NASA did plan for the shuttle to be launched from California into polar orbits, but after the *Challenger* accident, those plans were cancelled. There were two primary reasons for this cancellation. One was the fact that there were many more satellites waiting to be launched from Florida than California and NASA needed all three of the remaining shuttles to make these launches. The other reason was that California launches were going to require a much lighter solid rocket booster design. Remember, launching into polar orbits means you don't have the earth's free eastward rotational speed, so NASA needed a way of cutting weight from the boosters to minimize the reduction that otherwise was going to be required in payload weight. So, instead of using steel, they were going to make the boosters out of nonmetal composites similar to fiberglass. Having already experienced one tragedy as a result of a solid rocket booster failure, NASA was reluctant to take a chance on this completely new booster design. They scrubbed the plan to use the California launch site.

## Can the shuttle fly around the earth's equator?

No. A shuttle launching from the Kennedy Space Center (lattitute approximately 28.5 degrees) doesn't have enough fuel to bend its orbit southward to coincide with the equator (lattitude 0 degrees). The minimum inclination it can fly is the lattitude at the launch pad, or approximately 28.5 degrees latitude. This means astronauts on minimum-inclination missions won't get to see much of the earth's land masses. Their nadir will pass only over the area between 28.5 degrees north and south latitudes. In other words, the only part of the United States they'll pass over will be extreme south Texas, the southern half of Florida, and Hawaii. I hate to say low-inclination orbits are boring, but, compared with high-inclination orbits, they are. Unfortunately, astronauts don't get a choice on their inclination. As I explained earlier, it's determined by what you're going to do on the flight— the mission objective.

## How much payload weight can the shuttle carry?

Orbit inclination also affects how much payload weight a shuttle can carry into orbit. High-inclination missions aren't aligned with the spin of the earth, so more fuel (and less payload) has to be carried to make up for the lost free speed. NASA advertises that the shuttle can carry 59,600 pounds of payload into a minimum-inclination orbit (28.5 degrees) but only about 45,600 pounds into a high-inclination orbit (57 degrees).

## Why don't space-walking astronauts fall off the shuttle?

First of all, on most spacewalks, the astronauts are tethered to the shuttle, so they can't fall off. But even if they released their tethers they still wouldn't fall or be left behind, because they are traveling in exactly the same orbit and at exactly the same speed as the shuttle.

If you have a difficult time visualizing this, try this simple experiment. The next time you're a passenger in a car, driving at highway speeds—60 mph—hold a coin toward the car's ceiling in one hand and then drop it into the palm of your other hand. During the eighth of a second or so of drop time, the car travels approximately 11 feet (60 mph is 88 feet per second). Now, ask yourself, "When the coin was released, why didn't it slam into the back window as the car moved 11 feet forward?" The answer is that nothing changed the forward speed of the coin *relative* to the car. As the coin fell, it gained downward speed, but nothing changed its forward speed. It and the car continued forward at 60 mph. That's exactly the reason why a spacewalker continues in orbit with the shuttle. Her forward speed relative to the orbiter remains unchanged when she releases her tether.

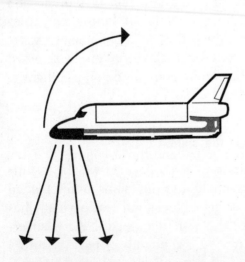

Downward-firing nose jets rotate the shuttle's nose up.

## If there is no air in space, how can astronauts steer a shuttle?

On Earth, when a pilot wants to steer his plane, he moves the stick, which in turn moves aileron and elevator controls to change the flow of air across the wings and tail and make the aircraft point where he wants it to point. In space there is no air, so it's impossible to use aircraft-like controls to change the direction a shuttle is pointed. Instead, the shuttle uses a system of 44 small rocket jets mounted in the nose and tail—the *reaction control system*, or RCS. These jets are pointed up, down, and sideways and are turned on and off in response to movement of the pilot's *rotational hand controller*, or RHC. Suppose an astronaut is docking her space shuttle with the Russian *Mir* space station. As she approaches, she needs to raise the nose. She will do this by tilting back on the RHC control stick, just as an aircraft pilot would do. But now, that stick action fires downward firing rockets in the nose to rotate the nose up. This is where the term *reaction* comes from.

When the jet fires downward, Newton's third law says there will be an equal and opposite reaction—equal magnitude in the opposite direction. In this case, the *reaction* is to raise the nose. To stop the rotation, the commander has to tilt the RHC stick forward. This will fire the upward-pointing jets in the nose and stop the movement.

Because the RCS rocket jets are in the nose and tail and because they point up, down and sideways, various jet firings can be commanded to point the shuttle (to rotate it) in any direction. Astronauts don't have to think about what jets need to fire, because the shuttle's computers will interpret the stick movements and fire the appropriate jets for them.

### Can the shuttle change its orbit?

Moving the RHC only changes the shuttle's *attitude*—its pitch, yaw, and roll. It will not change the shuttle's orbit (its altitude or inclination). To do this, you have to change the shuttle's speed. Large speed changes are done with the *orbital maneuvering system* (OMS). This system has two 6,000 pound thrust engines that can be fired only through the shuttle's computers.

Small speed changes, and thus small orbit changes, are done by firing various combinations of RCS jets. In space talk, such burns are called *translations* and are done through movement of the *translational hand controller* (THC). This is a stick diffrent from the RHC. The THC is a square-knobbed hand controller that can be moved up, down, left, right, and in and out. These movements cause the appropriate RCS jets to fire and move (translate) the shuttle in the corresponding direction. In other words, an inward movement on the THC fires the aft-pointing jets and moves the shuttle forward. An upward movement of the THC fires the down-pointing jets in the nose and tail and moves the shuttle upward.

There are RHCs and THCs for the commander (the person in the front, left seat) and on the back instrument panel. The set of controls in the back are used during rendezvous and docking. The commander will float at the aft cockpit with the RHC in his right hand and the THC in his left hand and fly the shuttle while watching the target out of the back and top windows.

Before leaving this RCS discussion, I should mention that every shuttle astronaut remembers the first time she experiences an RCS jet firing. This occurs during ascent after the main liquid engines are shut down and the empty external tank is jettisoned. You want to get away from that tank so it doesn't bang into the belly of the spacecraft and damage the heat tiles. So, after the tank is released, the autopilot fires the down-firing jets in the nose and tail to move the shuttle away. The reason I say every shuttle astronaut remembers this event is because of the noise and vibration of the RCS firing. It takes you by surprise. You've gotten used to everything being smooth and quiet during the last few minutes of your ascent. Then, there's a thunk in the cockpit when the external tank is separated. For a couple seconds it's quiet again. Then, BOOM! BOOM! BOOM! The nose jets sound like cannons. The cockpit shakes and vibrates. Welcome to the RCS.

You might wonder how you can hear the RCS jets. There's no air in space to transmit the noise. You hear the jets because the sound is transmitted through the aluminum structure of the shuttle and into the air of the cockpit and then to your ear.

### How can satellites be parked in orbit?

They can't. Anything that's stopped in space above the earth—parked—would fall straight down. Yet, the term *parked in orbit* is frequently used by NASA. What gives? When NASA refers to a parked satellite, they are referring to a satellite in *geostationary* orbit. Objects in circular orbit 22,300 miles above the earth's equator circle the earth once every 24 hours. This means that to an observer on the ground, the satellite doesn't appear to move. It appears parked in the sky. But, correctly said, its motion is synchronous with the earth's surface movement, or *geosynchronous*. (More correctly, a 22,300-mile-high circular, equatorial orbit should be called *geostationary*, but I will use the commonly accepted term for such an orbit—*geosynchronous* or *geosync*.)

This fact of orbit mechanics—that an object in orbit at 22,300 miles above the earth appears parked—has spawned a communications revolution. It enables someone on Earth to point a satellite dish at a *fixed* spot in the sky and get a TV, telephone, or other communication signal from space. Think of how difficult it would be for satellite TV reception if every backyard Earth station dish had to *track* a moving satellite. There would be a much smaller satellite TV market because of the complexity of that tracking effort. But to track a geosynchronous satellite, a dish can be pointed once and left alone. The motion of the earth itself is doing all the tracking—it matches the satellite's motion.

Besides communication satellites, there are other users of geosync orbit. The weather pictures you see on the evening TV news come from meteorologic satellites at geosync altitude. Also, the military has missile early-warning satellites parked in geosync. These stare down at Earth with infrared eyes that can see the heat of a rocket launch and would give our leaders a warning in case of an attack.

**Can geosynchronous satellites be placed anywhere above the earth?**

No. The only geosynchronous orbit is 22,300 miles above the earth's *equator*. A satellite in an orbit titled to the equator will *always* be moving relative to an observer on the ground. If the earth is spinning due east and a satellite is in an orbit that's inclined northeast-southwest, it's obviously not matching the earth's spin, so it can't appear stationary to someone on the ground. Knowing this, you can chuckle, as I do, whenever you see a news media report or read in a novel that the United States has spy satellites parked over Lebanon or Iran or Russia. Space physics says this is impossible.

You have evidence in your own neighborhood of the fact that the only geosync satellites are above the equator. Take a look at all the satellite dishes you pass in back yards and at cable TV stations. What pointing direction do they have in common? If you're in the northern hemisphere, they all have a *southerly* aim. Some may be pointed due south, others southeast, and still others southwest (it depends on what satellite they are pointed at), but (in the northern hemisphere) they all have a southerly component of aim because that's the direction to the equator. If you are in the southern hemisphere, you would see all satellite dishes pointed in a northerly direction.

### Is it possible to see an orbiting satellite from Earth with the naked eye?

Absolutely! I have seen hundreds of satellites from Earth. On rare occasions, I've even seen five or six satellites crossing the sky at the same instant. But they can be seen only when the sun is shining on them and it's dark on the earth. In other words, they are only visible beginning about an hour after sunset for about a 1-hour period or about 2 hours before sunrise for a 1-hour period. During these times, like sun shining on a mountain peak while it's still dark in the valley, the sun will shine on a satellite while it's dark on the earth directly below. This sun glint against the black of a dark sky is frequently bright enough to be seen—and probably accounts for a number of UFO sightings.

### What is the best way to watch for satellites from Earth?

First, try to get away from as much light as possible. It would have to be an exceptionally bright satellite to be seen from a city or well-

lighted suburban yard. Second, lie on a blanket or recline in a lounge. If you remain standing, you'll get a pain in your neck. Third, do not try to use binoculars or telescopes. Remember, you are trying to detect *motion* and that's very difficult to do in the narrow field of view of an opti-

**An observer in the dark will be able to see a sunlit satellite overhead.**

cal device. Instead, let your unaided eyes roam the sky, pausing briefly for a few seconds in different areas of the heavens. If there's a passing satellite, you'll see it moving against the background of the stars. But spend most of your search time in the hemisphere of the sky that's down sun, away from the setting or rising sun. You're looking for *reflected* light and you'll see this best when the satellite is presenting a whole side of illuminated surfaces to you. Fourth, don't confuse an airplane with a satellite. Airplanes will have flashing red and white lights and might even be leaving a vapor trail or making audible noise. Satellites have no lights on them and will not leave vapor trails or make noise. The only reason they are visible is because of reflected sunlight; therefore they will always appear whitish. Also, because of the changing angle of the sun's reflection as satellites move across the sky, they might vary in brightness, sometimes even going out for several seconds between reflections. They also usually appear to move faster than an airplane. Another difference between airplanes and satellites is that eastward orbiting spacecraft will suddenly go out and remain out as they enter the earth's shadow, while the lights of a moving plane will remain visible all the way across the sky. Finally, don't stare at the satellite once you see it. In darkness, our eyes have a natural blind spot at the focal point so if you stare at the satellite it'll seem to disappear. Just scan around it.

## Can you see an orbiting space shuttle?

Unless you have seen a prediction of a shuttle passage in a news-paper or other source, it would be impossible to know you are seeing a shuttle fly over. You won't be able to see its wings or other detail because of the distance. But it is possible to make some logical conclusions about the objects you are viewing. First, the faster they are moving, the lower their orbit. I won't go into the orbital mechanics of this, but it's a fact that higher orbits are slower orbits. It takes the moon a month to orbit the earth, while a shuttle takes only 90 minutes. In all cases you will be seeing satellites that are only a couple of hundred miles above the earth. It's impossible to see anything at geosynchronous orbit with the naked eye. Second, if the object varies significantly in brightness as it travels across the sky, it's probably a piece of space junk. Functional satellites are usually stabilized and will be less changeable in brightness because of that stability. A piece of junk will be tumbling, thus the changes in brightness. Third, notice the direction a satellite is traveling. If it's in a high-inclination orbit—steeply tilted to the equator—then smile. There's a good chance it's a military satellite and it might be taking pictures of you. Most of these are in orbits that are generally north-south in direction so they can see the majority of the earth as it spins underneath. If the satellite is crossing in a generally east-west direction, it's probably some astronomic observation science satellite (e.g., the Hubble Space Telescope) or a thrown-away rocket body used to launch a communication satellite.

For those of you interested in knowing when a space shuttle might be passing over your location, you might want to send an inquiry to this Internet address: adamod@iapc.net. This is the address for Dan Adamo, a NASA engineer who has prepared a home computer software program that shows the path of any orbiting space shuttle across the earth.

## Are orbit rendezvous dangerous?

I chuckle whenever the press hypes a shuttle rendezvous with particular emphasis on the incredible speed at which the rendezvous occurred—17,300 mph—the implication being this speed made the rendezvous exceptionally dangerous. Not so. The absolute speed of the shuttle (the 17,300 mph) has nothing to do with the danger involved. It's the relative speed *between* it and the object it's approaching that matters. As long as that relative speed (the difference in the speed of the objects) is nearly zero, a rendezvous of two spaceships traveling at 17,300 mph isn't any more dangerous than a rendezvous of two aircraft traveling at 173 mph. In fact, some pilots would tell you that a nighttime rendezvous of two aircraft in marginal weather is a lot more stressful than an orbit rendezvous.

## Are orbit rendezvous difficult?

Yes. Imagine you are the commander of a shuttle on a mission to rendezvous with the Russian *Mir* space station. It's at your exact altitude and 20 miles in front of your shuttle. You need to gain on it—close that 20-mile distance—so you do what's intuitive. You push in on your THC, firing the aft thrusters, thus increasing your speed. Everything you've ever done on Earth tells you this increase in forward speed will bring you closer to *Mir*. But in orbit, it actually has the *opposite* effect. Within minutes, you'll find the distance to *Mir* increasing not decreasing. What's going on? Increasing the speed of any orbiting spacecraft will raise the orbit, and, by the laws of orbit motion, higher orbits are slower orbits. So, your forward burn raises the shuttle orbit and causes it to slow down. The distance to *Mir* increases. To close the distance, you need to do a *braking burn*. This will lower the shuttle orbit and increase its orbit speed. Eventually, through a series of burns, the distance will be closed and the rendezvous completed. Only when the shuttle is within about 2 miles of a target could a commander complete a rendezvous by eyeball. Beyond this distance, the counterintuitive behavior of orbit mechanics makes him a slave to radar and computer data.

# CHAPTER 2

# Space Shuttle Pre-Mission and Launch Operations

## Why did we build a space shuttle?

To answer this question, think back on what America used to put astronauts and satellites in space before the shuttle: Delta, Atlas, Titan, and Saturn rockets. All of these were throwaway rockets. Nothing was used again. The rocket stages were jettisoned into the ocean. Even the capsules that contained the astronauts were never used again. They were sent to museums. The shuttle was designed to eliminate this waste. Think of it as the first recyclable rocket. By using parts over and over, it was thought that the cost of launching satellites could be substantially reduced. Everything on the shuttle is reused except the big, orange gas tank. Nearing orbit, that's jettisoned to burn up in the atmosphere.

## How big is the ready-to-launch space shuttle system?

At lift-off the entire stack (that's NASA slang for everything stacked on the pad and ready to launch) weighs about 4.5 million pounds and stands 184 feet from bottom to top. Approximately 4 million pounds of this weight is liquid and solid fuel.

To get this off the ground, the shuttle's liquid- and solid-fueled engines generate about 7.5 million pounds of thrust. These numbers sound impressive, but the shuttle is a toy compared with the rocket that took men to the moon—the *Saturn V*. It stood another 150 feet taller than the shuttle and was nearly 3 million pounds heavier.

## What are the main components of the space shuttle?

The space shuttle stack is comprised of three major pieces: the orbiter (the winged spacecraft that carries the as-

**Shuttle stack is carried to launch pad.**

tronauts and payload), the solid rocket boosters (SRBs) (the white tubes) and the external fuel tank (the orange belly tank).

### How big is the orbiter?

The orbiter is about 122 feet from nose to tail, 78 feet from wing tip to wing tip and weighs about 173,000 pounds (when empty of maneuvering fuel and payload). Three liquid-fueled engines attached to its rear each produce nearly 400,000 pounds of thrust. Its payload bay is 60 feet long and 15 feet in diameter, and it can carry a maximum payload of 60,000 pounds into a low-inclination orbit 150 miles above the earth. A completely reusable thermal protection system composed of about 28,000 tiles and blankets protect the orbiter from the nearly 2,500 degree heat of reentry air friction.

### How big is the external fuel tank?

The *external tank*, or ET, is 154 feet long and 27.5 feet in diameter and carries approximately 1.6 million pounds of liquid oxygen and liquid hydrogen, which powers the orbiter's liquid-fueled engines. Large openings in the belly of the orbiter allow fuel lines to carry the fuel to the engines. After the ET is jettisoned, doors are closed over these openings to form a perfectly flat belly that can withstand the heat of reentry. The ET is the only part of the shuttle system that is not reused. After being jettisoned, most of its 66,000 pounds of empty mass will burn up in the atmosphere. The pieces that survive will impact in remote areas of the Pacific ocean.

### How big are the solid rocket boosters?

The SRBs are attached to the sides of the ET. Each is about 150 feet tall, weighs 1.3 million pounds, is over 12 feet in diameter, and produces 3.3 million pounds of thrust. Their huge size makes it impossible for the boosters to be manufactured and shipped in one piece, so they are segmented. Each is made up of four propellant-filled pieces that are shipped separately by rail from the factory in Utah to the Kennedy Space Center where they are bolted together to form one complete rocket. The SRBs are ignited at T minus 0 (launch time minus zero, or lift-off) and burn for 2 minutes and 11 seconds to help push the shuttle to about 25 miles altitude and to a speed of approximately 3,000 mph. At burn out, they are jettisoned and parachute into the water 170 miles from the launch pad. (Even though they use the world's largest parachutes, they still hit the water at over 100 mph.) Tug boats recover them and tow them back to shore for reuse. Each weighs about 190,000 pounds when burned out, but they float because air is trapped inside them.

### Why do some shuttle photos show a white external fuel tank instead of an orange one?

The shuttle engines can only burn *liquid* oxygen and *liquid* hydrogen, but these propellants will boil into an unusable vapor at minus 298° F and minus 423° F, respectively. So, the external fuel tank must be insulated like a giant thermos bottle to keep the propellants cold enough to remain liquid. A foam is sprayed onto the aluminum tank to provide this insulation. The natural color of this material is burnt orange. For the first two shuttle missions, however, the insulation was painted white because it looked better. NASA later decided such aesthetic considerations were too expensive to continue. Besides the labor costs involved, the paint added several hundred pounds of useless weight to the tank. Every astronaut would rather have extra fuel and payload than a nice, white ET.

### How do astronauts train for launch, orbit, and reentry?

At the Johnson Space Center in Houston, Texas, NASA has two major shuttle flight simulators: the *motion based simulator* or MBS and the *fixed base simulator* or FBS. Both have cockpits that are identical to the shuttle, and all the switches and displays work just as they would in a real shuttle. Basically, they are like giant video games that use computers to duplicate how instruments and displays in the cockpit and in Mission Control would appear during a nominal mission and with various shuttle system failures.

During simulator training, instructors sit at controls where they can fail various pieces of equipment and see how the crew and mission controllers respond to these failures. The training is brutal—particularly launch training. It's during ascent that things happen so fast any mistake could be disastrous, so astronauts and mission controllers are really stressed with failures during simulated launches. The instructors might simulate an engine failure at 1 minute into flight, a hydraulic failure at 1 minute 30 seconds, an electrical failure at 2 minutes, a navigation failure at 2 minutes 30 seconds, a computer failure at 3 minutes, a fuel leak at 3 minutes 30 seconds, and so on. The individuals who prepare these training scenarios are called *sim sups*, for *simulator supervisors*. Astronauts joke that sim sups wear starched underwear and shoes that are two sizes too small so they can make themselves as mean as snakes and dream up all manner of diabolical failures. On the astronaut training schedule, launch simulations are listed as "Ascent Skills," but, because of the sim sups' creative torture, astronauts jokingly call this training "Ascent Kills."

There is one big difference between the simulators, however. The FBS is, as the name implies, fixed. It doesn't move and is used primarily for orbit operations training. The cockpit of the MBS, however, does move and it's used for launch and reentry training. It sits on six hydraulic legs, looking like a giant spider. A computer controls these legs and they extend and contract to make the cockpit shake, tilt, and roll to give the astronauts a little taste of what a real launch and reentry feel like.

### Does the training program really prepare you for spaceflight, or are there surprises?

NASA's astronaut and mission controller training programs are so thorough it's hard to imagine anybody could ever be surprised by any equipment failures. Before launch, every imaginable failure mode will have been practiced countless times, and astronauts' reactions to emergencies become as automatic as breathing. For example, on my first mission, during the middle of a congratulatory call from President Ronald Reagan, an alarm sounded. Did we cut off the President and tell him to wait while we investigated a malfunction? Not a chance. We never broke stride with the call and simultaneously troubleshot the problem. Somebody grabbed the malfunction checklist. Somebody punched up the correct display on the computer screen. Somebody floated to the panel that controlled the failing system. And throughout this response, the microphone was passed around and around while we all said, "Yes, Mr. President" and "Thank you, Mr. President" and "Everything is going great, Mr. President." The problem turned out to be very minor. A heater on a piece of shuttle equipment had failed, and we merely switched to the backup heaters.

The only thing the NASA simulators are deficient at duplicating are the *physical* sensations of a mission. The noise, vibrations, and g-force of a simulated launch do not compare with the real thing. It was this deficiency that brought me my biggest surprise. On my first mission, at SRB separation, the dramatic drop in noise and vibration shocked me. In fact, I had a momentary fear that even our three liquid engines had shut down. And *standing* around the FBS for an orbit simulation is obviously not the same as *floating* around in a real cockpit.

### How do astronauts train to land the shuttle?

The MBS and FBS are great for instrument and procedure training, but at the end of every mission the commander is going to have only one chance to land a 100-ton glider that's diving at 300 mph. Is there someway to prepare for this task? Yes. Commanders and pilots train for the final 30,000 feet of a shuttle descent and landing with the *shuttle training aircraft*, or STA—a Gulfstream business jet that uses computers to make it glide like a real shuttle. (The engines have to be reversed in flight and the loading gear and flaps lowered to make the jet sink like a rock, as a real shuttle does.) The left side of the STA cockpit has an instrument panel, stick, and speedbrake controls like a real shuttle, and inserts are put in the windows to duplicate the view a commander would have from the shuttle cockpit. An instructor pilot, sitting in the right seat, flies the STA to about 30,000 feet altitude and then gives control to the astronaut in the left seat to fly to a simulated shuttle landing. Before an astronaut can serve as a shuttle commander, she must have done a minimum of 800 practice STA landings. No wonder all those landings you see on TV look so smooth.

### How do astronauts train to use the robot arm?

Can you imagine lifting a satellite from the shuttle payload bay that's the size of two Greyhound buses, costs several hundred million dollars, and is fragile as an egg? To compound the difficulty, of the job, imagine having only inches of clearance between the satellite and the shuttle structure. NASA prepares astronauts for this challenge in

*Helium balloons simulate weightless payloads in robot arm training.*

the *manipulator development facility*, or MDF. It's a full-scale mockup of the shuttle payload bay, with a simulated robot arm attached to the left side. Giant, helium-filled balloons simulate weightless payloads, and astronauts practice raising and lowering these to prepare for robot arm operations in orbit.

### Do astronauts train for mission g-forces in a centrifuge?

A centrifuge is a spinning arm that uses centrifugal force to create a higher g-force at its end. Some readers might have actually ridden in something similar to a centrifuge, because some carnival rides have spinning chambers that use centrifugal force to pin you to the wall while the floor is moved from under your feet. Rookie astronauts ride in an Air Force centrifuge that simulates the g-force profile of a shuttle launch: 1.6 g's at lift-off, slowly building to 2.5 g's by 2 minutes, dropping to around 0.9 g at SRB burnout, then slowly building to 3 g's nearing main engine cutoff.

### Do shuttle astronauts train in a multi-axis chair?

Many people have seen photos and movies of astronauts sitting at the center of a gyroscopic-like device and being wildly tumbled in many different directions simultaneously. Space Camp and some carnival rides have these multi-axis chairs. Actually, shuttle astronauts do not train in these devices. They are simulations that were used in the early space program when astronauts traveled in small capsules. During that time, NASA engineers were worried that a capsule could have a steering malfunction that could spin it into a complex tumble. They built the multi-axis chair as a tool for astronauts to practice recovering from such a malfunction. The chair had controls that simulated the capsule steering jet firings and enabled the astronauts to fly out of the tumble (i.e., to regain control of the capsule).

### Do astronauts train in an antigravity room?

When I'm asked this question, I'm reminded of a story I heard at Space Camp. During a Spacelab briefing for a group of young campers, counselors were emphasizing the need for the kids to use the foot loops inside the lab to restrain their bodies and to Velcro all items, because in space nothing can be put down. Everything will float. Afterward, they directed the kids to enter the Spacelab mock-up. As the hatch was opened, one child took a running, superman style leap through it and immediately crashed onto the Lab floor. The incredulous counselors ran to him to see why he had done such a crazy stunt. "You forgot to turn on the antigravity machine!" was his angry response.

Unfortunately, there is no such device as an antigravity machine or an antigravity room. Photos you might have seen of astronauts floating in a room were taken aboard NASA's Vomit Comet. Remember, you can never turn off gravity or get away from it. The only way you can be weightless is to be freely under the influence of gravity's pull.

### How long does it take to prepare a just-landed shuttle for another mission?

After a shuttle lands, it is towed into a hanger-like facility called the *orbiter processing facility*, or OPF, where it will remain for about 2 months as it's prepared for its next mission. The payload support equipment from its last mission has to be removed and the new payload support equipment installed. Numerous inspections of the orbiter systems have to be made. Engines have to be serviced or changed. There might be a requirement to do a modification or repair some damaged heat tile or perform some other unique maintenance. Eventually, the vehicle is moved to the *vertical assembly building*, or VAB, where it is stacked with the SRBs and the external fuel tank. This process requires about a week. The entire stack is then moved to the launch pad where more tests are done, the payload is loaded, fuel is loaded, and the rocket is finally launched. This pad time averages about 1 month, meaning the average orbiter turnaround time from landing to launch is approximately 90 days.

### How many shuttle missions does NASA launch each year?

NASA averages seven to eight shuttle missions per year, but since the beginning of the shuttle program in 1981, the number of missions per year has ranged from as few as none to as many as nine. Here's a year-by-year breakdown:

| | | | | |
|---|---|---|---|---|
| 1981—2 | 1984—5 | 1987—0 | 1990—7 | 1993—7 |
| 1982—3 | 1985—9 | 1988—2 | 1991—6 | 1994—7 |
| 1983—4 | 1986—2* | 1989—5 | 1992—8 | 1995—7 |

*(including the *Challenger* tragedy)

**How many shuttle missions have there been?**

As of November, 1995, NASA had launched 72 shuttle missions. Missions by orbiter vehicle are as follows:

| | |
|---|---|
| *Columbia:* | 18 |
| *Challenger:* | 10 (including STS-51L, the flight that destroyed *Challenger*) |
| *Discovery:* | 21 |
| *Atlantis:* | 14 |
| *Endeavour:* | 9 |

**Why doesn't the shuttle *Enterprise* ever fly in space?**

Six shuttles have been built: *Enterprise, Columbia, Challenger, Discovery, Atlantis,* and *Endeavour. Challenger* was destroyed by an in-flight explosion in 1986, and *Endeavour* was built to replace it.

Â *Enterprise* never flies in space because it was never designed for spaceflight. It was built only to test the shuttle's atmospheric gliding abilities. On five occasions, before the first shuttle was launched, it was carried on the back of a Boeing 747 and released at about 30,000 feet and flown to a landing at Edwards Air Force Base. It may have looked like all the other shuttles on the outside, but on the inside it was significantly different. Many of the systems needed to fly in space were never installed.

Â *Enterprise* was retired after its very brief air-drop career and is now the property of the Smithsonian Institution. Ultimately, it will be placed in a hanger at Dulles airport in Washington, D.C., as a Smithsonian museum display.

Â Blame "Star Trek" fans for the fact that there will never be a shuttle *Enterprise* in space. Originally, NASA did not intend to give the glide-test orbiter the name *Enterprise* but when the Trekkies heard this, they mounted a campaign to get NASA to use the *Enterprise* name on the first shuttle. NASA caved in to their demands. If the Trekkies had done their homework, they would have found that the first shuttle was to be the glide-test orbiter and would never fly in space.

So there are now four flying shuttles: *Columbia, Discovery, Atlantis* and *Endeavour*. In all likelihood there will be no others. The assembly line has been closed for several years.

Except for *Columbia*, which is about 8,000 pounds heavier than the other orbiters, they are essentially identical in construction. *Columbia* was built heavier because it was the first to fly, and engineers were uncertain of all the stresses it might have to withstand. With the data from her early missions, they were able to make her sister orbiters lighter.

## Who names the shuttles?

The NASA Administrator has the authority to name NASA spacecraft. After *Challenger* was lost, NASA solicited suggestions from school children to choose the name of the replacement shuttle, *Endeavour*.

On the countdown for my second launch (aboard *Atlantis*), my commander, Hoot Gibson, had us laughing with this comment about shuttle names, "Do you guys realize we're launching in a rocket named after a sunken continent?" He went on to say, "If you're going to name a shuttle *Atlantis*, why not *Titanic* or *Pompeii* or *Krakatau* or even *Sodom & Gomorrah?*" He had a point.

## What do the shuttle mission numbers mean?

If you have an autographed copy of this book, you'll probably note the following cryptic letters and numbers under my name: STS 41D, 27, 36. (These are not my wife's measurements, as I've been asked.) These are my three shuttle missions, with STS meaning Space Transportation System, NASA's nomenclature for the space shuttle. The numbers and letters that follow represent the 12th, 27th, and 34th (not 36th) shuttle missions. It takes a little history to understand this apparently bizarre mission designation system.

When the shuttle program first started, NASA used a straightforward numbering system. The first shuttle launched was STS-1, the second was STS-2, and so forth. This continued until 1984. At that point, the shuttle launch manifest, which is prepared years in advance, began to reflect future shuttle launches from Vandenberg Air Force Base in California. (In 1986, after the *Challenger*, NASA cancelled plans to fly shuttles from California, so these missions never occurred. But in 1984, they were being planned.) Planners wanted the mission numbers to differentiate between Kennedy Space Center and Vandenberg launches, so they concocted a number and letter system. STS-12 (my first mission) became STS-41D. *Challenger*, the 25th shuttle flight, carried a designation STS-51L. The first digit of this new system indicated the planned fiscal year of the launch. The second digit—either a 1 or a 2—indicated the launch site. A 1 specified the Kennedy Space Center and a 2 a Vandenberg launch. The letter designation that followed the digits reflected the planned order of the mission in the fiscal year. An A indicated the first planned launch, a B the second launch of the year, and so forth.

With this explanation, you can now decipher my first mission, STS-41D. It was originally schedule to fly in fiscal 1984 (4) from Kennedy Space Center (1) as the 4th planned mission (D) of the fiscal year. The *Challenger* nomenclature, STS-51L, says it was originally placed on the manifest to fly in fiscal year 1985 (5) from Kennedy Space Center (1) as the 12th planned mission (L) of the fiscal year.

An important tidbit about this mission nomenclature is that it only reflected the *planned* launch manifest, not the *actual* launch history. For example, going back to the *Challenger* nomenclature, STS-51L, we see it was originally planned—many years in advance—to be the 12th mission of fiscal year 1985 from Kennedy Space Center. In fact, because of delays in its payload preparation, it ended up being the 4th mission in fiscal year 1986. Since it would have been a monumental task to have constantly kept all the shuttle mission paperwork updated to reflect the reality of launch delays, NASA adopted the sensible rule that once a mission was given a number, that designation would remain with it forever.

After the *Challenger* tragedy, plans to launch shuttles from Vandenberg Air Force Base were cancelled and NASA returned to the old system—STS-26, STS-27, STS-28, and so on. But they continue to assign the numbers permanently, based on the planned launch schedule. This is the reason you cannot be certain that STS-65 was the 65th mission or that STS-36 was the 36th mission. In fact, as I said earlier, STS-36 wasn't the 36th mission. Because of delays with other missions that had been planned to precede it, it was the 34th mission.

### Who designs the mission patches?

The crew is responsible for designing its own mission patch. If someone on the crew has some artistic talent, he might actually do the designing. Usually, though, astronauts are so busy and artistic talent is in such short supply, they will avail themselves of the many artists around the country who volunteer to do a mission patch. Sometimes the final results are eye catching, while other times a patch turns out to be an eyesore. The latter is usually the case when astronauts insist on including some aspect of the entire mission on the patch—for example, a robot arm grabbing a satellite with several spacewalkers hovering nearby with the earth as a backdrop and a solar eclipse in progress. All of this crammed onto a 4-inch-diameter patch does not make a pretty picture.

### Can space shuttles fly to the moon?

No. The moon is about 240,000 miles away. The space shuttle can fly to a maximum orbit of approximately 400 miles altitude. The only rocket that was ever capable of carrying people to the moon was the Saturn V, and NASA no longer makes that rocket. Whenever NASA next returns to the moon, you can be certain the rocket will look nothing like a shuttle. Because the moon has no atmosphere, wings—as the shuttle has—would be useless.

## Where does the shuttle go and what does it do?

As the name implies, the intended mission of the shuttle was to be a reusable spacecraft that would shuttle people, equipment, and satellites into orbit and to service and/or return satellites already in orbit—for example the Hubble space telescope. It was never intended to go to the moon or Mars or to be a long-term space laboratory. The shuttle will be the primary vehicle for shuttling the equipment and crews needed to build the space station. (Construction is planned to start in late 1997 or early 1998). After the space station is completed (in 2002), the shuttle will continue to support station operations by ferrying astronauts and equipment to and from it. When it's not engaged in these space station missions, it will continue to be used as a short-term science laboratory and to launch, service, and retrieve satellites. All shuttle missions fly into orbits between approximately 100 and 400 miles altitiude at inclinations (tilt) to the equator between 28.5° and 62.0°.

## What does a shuttle mission cost?

According to NASA, the average cost of a shuttle launch is about $470 million, paid by you, the American taxpayer. Before the *Challenger* tragedy, when the shuttle was being used to launch commercial communication satellites, those commercial enterprises were paying NASA for their launch services.

In any discussion about the cost of shuttle launches, you have to ask whether they are worth it. I certainly believe so, though others are very vocal critics of the program and its expense. The one thing you must keep in mind when considering the potential of space-based research is this fact: Space provides an environment that is *absolutely unique*. Prolonged weightlessness cannot be duplicated anywhere on Earth. Now, ask yourself, is it possible that some molecule could be joined with some other molecule in an orbiting laboratory to form an entirely new and beneficial drug that's impossible to manufacture on Earth? Is it possible that the cure for AIDS or cancer or diabetes or Alzheimer's disease could be developed in space? Is it possible that a new alloy could be produced in space that would revolutionize automobiles or airplanes or artificial hearts, lungs, or kidneys? Or, is it possible that you can make a computer chip in weightlessness that will start a new revolution in that industry to the benefit of our terrestrial economy? I can't answer those questions. In fact, nobody can—not NASA's critics or its supporters. That's the nature of research. You can't know until you do it. You can only say that the *potential* for breakthroughs in pharmaceuticals, alloys, and so furth, exists in the uniqueness of a weightless laboratory. So, when someone says to me that space research is too expensive, I have to ask them, "Too expensive compared to what outcome?" Is the cure for cancer worth a $30 billion investment in a space station? I think most people would clearly answer *yes* to that question. Simply put, we can't take a chance that we'll miss some incredible breakthroughs in an environment that is not possible to replicate on Earth—prolonged weightlessness.

### How many people does a space shuttle carry and what are their crew positions?

There is always a minimum of five NASA crewmembers aboard a shuttle: commander, pilot, and mission specialists 1, 2, and 3. Additionally, there may be one or two payload specialists aboard. The largest crew size ever flown in the shuttle was eight people. Theoretically, there is room for 10 people: four seats upstairs and six seats downstairs.

### What are a shuttle commander's duties?

The commander is, as the name says, the overall commander of the mission—like the captain of a ship. (Some astronauts have wondered if international law would allow a shuttle commander to officiate at a marriage in space.) During launch and landing, the commander sits in the front left seat and is the one to fly the shuttle to a nominal or an abort landing. He also manually flies the shuttle to a rendezvous/docking with other spacecraft from the aft cockpit. In other words, he's going to get all the real stick time. He will be well trained on all the shuttle systems, but particularly on the *data processing system* (DPS) (the computer system), and the *environmental control and life-support sub-system* (ECLSS) (heating, cooling, oxygen control, etc.). The controls for these systems are on his side of the cockpit. He will also have some training in the payload activities and will help the MSs in those activities.

### What are a shuttle pilot's duties?

The pilot sits in the right front seat during launch and landing—like a copilot on a regular airplane. She is as well trained as the commander to manually fly the shuttle to a landing or a rendezvous/docking with other spacecraft, but, in all likelihood, will only get a few seconds of real stick time. You have to remember, there aren't many times when *anybody* has their hand on the stick and during these times, things are happening so fast that the commander can't afford to be a nice guy and share the stick with the pilot. The pilot has to wait until she is a commander to really "fly" the shuttle. After one or two flights in the pilot position, most pilots get promoted to commanders. Like the commander, the pilot will have some training on the payload activities and will assist the mission specialists (MSs) in those activities.

On the pilot side of the cockpit are the controls for the *auxiliary power units* (APUs) (hydraulic system), the *reaction control system* (RCS) (steering jets), the *orbital maneuvering system* (OMS) (orbit adjustment engines), the *electrical power system* (EPS), and the *main propulsion system* (MPS) (liquid main engines). The pilot is trained to be a specialist in these systems, and during simulations, the pilot position is usually the hot box. Instructors will input multiple failures until the pilot is buried in life-threatening emergencies. No crew position is more stressful during launch and landing than the pilot's because of her control over so many critical systems.

## What are the mission specialists' duties?

Mission Specialists are the astronauts responsible for the orbit activities. They are designated MS1, MS2, and MS3. As the name implies, they are the specialists in whatever the mission is intended to do. If the mission involves releasing and/or retrieving a satellite, an MS is at the controls of the robot arm. If a spacewalk is required, MSs do it. Usually all payloads and experiments are the primary responsibility of the MSs. Note that the term *specialist* has nothing to do with the MS's university degree or pre-NASA job. It only refers to their shuttle mission specialty and can vary from mission to mission. So, on one mission an astronomer astronaut might be a specialist with the robot arm, and on the next he might be doing a spacewalk. On another, a medical doctor astronaut might be a specialist on a cardiovascular experiment, while on her next flight she'll be helping to release a science satellite.

The MS2 astronaut also has the additional duty of being a launch and landing flight engineer. She is the one that sits behind and between the commander and pilot and helps them to respond to emergencies. The commander decides which MSs will be responsible for which jobs.

Two MSs will always be trained for spacewalks, even if no spacewalk is planned. This is because there could always be an emergency that requires a spacewalk (e.g., if the payload bay doors fail to close).

There is a lot of good-natured joking between the commanders, pilots, and the MSs. Generally, MSs think of themselves as the brains behind the mission and the two front-seaters as trained monkeys. Of course, the commanders and pilots joke that the mission specialists are space geeks.

## What are payload specialists?

Payload specialists, or PSs, are not career NASA astronauts. They are people who have unique talents that are needed for some specialized space research. Usually they are scientists. For example, Drew Gaffney was a heart specialist who flew on STS-40 to do weightless cardiovascular studies. Payload specialists receive minimal training on shuttle systems (toilet use, food preparation, emergency escape, etc.). After flight, they return to their universities or corporations and continue their research.

The term *payload* in their title is intended to reflect that they are not *shuttle* specialists as are the other career astronauts. They are aboard only to operate their own, unique payload experiment. This may or may not be part of the primary mission payload that the other career astronauts oversee.

### How many switches are in a shuttle cockpit?

There are over 1,000 hardware and software switches, controls, circuit breakers, and computer displays. No wonder astronauts spend thousands of hours in simulators to become experts on the shuttle systems. Even visiting cosmonauts are amazed at the level of control astronauts are given over the machine they ride. In fact, if a shuttle crew lost all contact with Mission Control immediately after lift-off, they could still fly into orbit and return safely to Earth (though they would probably use up ten lifetimes' worth of adrenaline in the process).

The first time I climbed into a shuttle cockpit, I wasn't surprised by the number of switches as much as I was by the physical *size* of the switches. I expected weight would be a critical factor in the design of a spaceship, so I envisioned that everything—switches included—would be minimized in size and weight. Instead, the switch knobs, buttons, and toggles look as if they were designed for a battleship. They are large. But I quickly learned why. They have to be functional for an operator wearing bulky pressure suit gloves. Imagine having to operate your TV remote control with ski gloves on your hands. Obviously, in such a hypothetical situation, you would want the remote control buttons to be huge. So it is on the shuttle.

## What do astronauts wear when they launch into space?

On launch morning, the first thing an astronaut dons is a *urine collection device* (UCD). For women, this is an adult-style diaper. Men have a choice of a diaper or a condom-type device that Velcros around the waist.

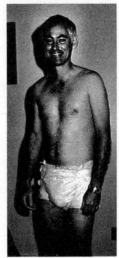

Next comes a set of Patagonia long underwear and boot socks. At this point, the astronauts walk from the crew quarters and down a hall to the Suit Room. Here, technicians help dress them in the familiar orange pressure suit and boots. This suit serves two emergency functions. First, it's designed to protect astronauts in the event of a cabin pressure leak. If such a depressurization ever occurred and a crewmember wasn't wearing a pressure suit, the gases in his blood would come out of solution (his blood would boil) and he would be killed. In such a situation, a pressure suit (it's basically a rubber suit) would automatically inflate with air and keep enough air pressure on his body to keep

**Author in a urine collection device.**

him alive long enough to return to Earth. The pressure suit's second emergency function is to serve as an anti-exposure suit in the event the crew had to bail out. Such thermal protection is essential to survive in cold ocean water.

After dressing in the pressure suit and boots, astronauts leave for the launch pad. The next items of apparel are added at the White Room—a small room just outside the shuttle's entry hatch that is painted white to make any dirt more visible for clean-up. Here technicians help each atronaut into a parachute harness, pressure suit helmet, and lumbar pad. This last device is an inflatable air bladder that is Velcroed to the lower back of the suit. A rubber ball—similar to the ball that inflates a blood pressure cuff—extends around the side of the suit so the astronaut can hand-pump the pad with air in an attempt to alleviate the back pain associated with long waits on the launch pad. (It's marginally effective.) Next, the astronauts crawl through the hatch and into their seats, where they are attached to their parachutes and survival kits (these include a life raft).

*Author in a launch/entry pressure suit.*

The total weight of everything an astronaut is wearing while strapped into the seats for launch and reentry is about 83 pounds. In other words, in the event of a bailout emergency, this is the weight an astronaut has to carry to the hatch. For some of the petite female astronauts, this represents 80% of their body weight—the equivalent of my carrying 130 pounds. (Anybody who says women are the weaker gender hasn't seen female astronauts in shuttle bailout training.)

## Why did some astronaut crews wear sky-blue coveralls to launch while others wore orange suits?

The first four space shuttle missions were considered test flights and the two-person crews wore orange pressure suits to protect them in the event of a cockpit depressurization. In what *Challenger* later proved to be a significant mistake, NASA declared the test phase over after just four missions, and STS-5 became the first operational flight. The term operational implied that the shuttle had a reliability comparable to an airliner and, therefore, continuing to burden the crew with pressure suits was considered unnecessary. So, beginning with STS-5, astronauts launched in sky-blue coveralls. After the *Challenger* exploded, NASA returned to providing the crew with the protection of a pressure suit. In photos of all missions launched since *Challenger*, the crew is seen boarding the astro-van in these bulky orange suits.

**Does Mission Control monitor the astronauts' heart rate on launch?**

No. Unless it's part of a medical experiment, shuttle astronauts do not have a heart monitor or other biomonitors on their bodies at launch. In fact, the only time during nominal mission operations that astronauts wear heart-monitoring equipment is during spacewalks. This information is needed to ensure astronauts don't exert themselves beyond the cooling capability of their spacesuits.

**How do astronauts get in their seats when the rocket is on its tail?**

When the shuttle is on the launch pad, the back instrument panel is the floor. Obviously, you don't want to be standing on these switches when you are climbing aboard the vehicle, so platforms are installed to cover these controls. Technicians help the astronauts into their seats, and then remove these platforms as they exit the cockpit. In an emergency launch pad escape, astronauts would stand on and crawl over the switch panels.

**Do the astronauts have form-fitted seats?**

Many people think astronauts are waiting for launch in nice, comfortable chairs or "Star Trek"- type office furniture. Actually, the shuttle seats are just two pieces of flat steel with a thin cushion. They are miserably uncomfortable, and the intercom is frequently blue with curses—particularly when the launch is delayed for long periods. (NASA says the uncomfortable, flat-steel seat design is necessary to withstand the g-forces of a crash landing.)

**Does NASA schedule shuttle launches around female astronauts' menstrual cycles?**

No, and this part of female physiology has never been an issue with any part of shuttle operations.

## Have you ever had a mission scrubbed?

Have I ever! I never launched on any of my three missions on the first attempt. In fact, I strapped in a total of nine times to fly three times. On the six other attempts, bad weather or equipment malfunctions resulted in a scrub.

## Why does NASA delay the launch if it's cloudy? Can't the shuttle launch through clouds?

When NASA is launching a shuttle, it has four major weather concerns: lightning, wind, rain, and visibility. There have been several incidents of rockets (not the shuttle) being hit by lightning during launch. The long stream of exhaust coming out of the boosters tends to act like a lightning rod and attract lightning. In fact, several unmanned rockets have been lost because of lightning strikes. So, NASA has instruments that measure the static electricity in the air and will delay the launch if it appears lightning could be present.

Wind is another concern, because a very strong wind from the wrong direction could blow a launching shuttle into the launch support structure. Also, in an emergency landing (it's called a *return to launch site* [RTLS] abort), the wind could be a problem. If it's blowing too strong cross-wise to the runway, it puts extra strain on the landing gear and tires and might cause the shuttle to go off the runway. Even though an RTLS abort is unlikely after launch, to be extra safe NASA requires the wind at launch to be within limits for an emergency landing.

Rain is a significant problem. While the shuttle heat tiles are very good at protecting the craft from the 2,500 degrees of heat during reentry, they are otherwise very fragile. In fact, if you could hold one, you would think you had been handed a piece of styrofoam. The material is very light, and it would be easy to stick your finger into one. Just imagine the damage that could be done by a trillion objects hitting the heat tile at 600 mph— exactly the situation if a shuttle ascended into a rain cloud. Each raindrop would become a 600-mph bullet. The tiles would be eroded as if they were made of clay. That's not going to affect the shuttle's ascent, but it would put the crew in danger when they ultimately had to come out of orbit. Damaged tile could fail to protect the aluminum skin of the shuttle from the tremendous heat of reentry, and the shuttle could burn up like a shooting star—all because of launching through some rain.

It would seem that ground visibility wouldn't be an issue because the shuttle is headed into space. Who cares if it's foggy or cloudy beneath you? If you could be absolutely certain you were going to make it to orbit, Kennedy Space Center visibility wouldn't be an issue. But, in an RTLS launch emergency, the shuttle would have to turn around in flight and land at the Kennedy Space Center. Obviously, in this scenario visibility is going to be important. The commander wants to be able to see where she is landing. The shuttle has a zero-visibility, auto-land capability, and theoretically the crew doesn't need to see the runway to land, but why take the chance? It would be safer to delay the launch until the landing visibility is better. Also, the failure necessitating the emergency return might also result in a failure of the auto-land systems, so the pilot would have to be able to see the runway to land normally.

Finally, it could be completely clear with calm winds in Florida and the launch might still be delayed for bad weather conditions at emergency landing sites in Africa and Europe (a *Trans-Atlantic launch* [TAL] abort would have the shuttle landing at airports in Africa or Europe).

### If the shuttle blew up on the launch pad, would the astronauts be able to escape?

If the shuttle unexpectedly blew up on the launch pad, there would be no hope that a crew, strapped into the cockpit, could escape. However, there are other life-threatening launch pad emergencies in which the crew might be able to reach safety. For example, a leak of our hypergolic propellant (used in the OMS

**Astronauts in launch pad escape baskets.**

and RCS systems) would be a very serious emergency. Hypergolic fuels are extremely toxic. A leak wouldn't necessarily cause the rocket to blow up, but the fumes could kill an astronaut. In this, or other emergency situations in which the astronauts have some time to react, the crew would want to get away from the pad as quickly as possible. Taking the elevator or stairs wouldn't be very smart because the base of the rocket (where the stairs and elevator lead) may be the point of greatest danger. To provide astronauts with a fast way of escaping the launch pad area, NASA has installed escape baskets at the cockpit level of the gantry.

These baskets (there are seven of them) are attached to rollers that ride individual steel cables stretching 1,200 feet horizontally from the launch pad to the ground. Escaping astronauts (without any help from rescue crews) would unstrap from their seats, crawl to the side hatch, open it, run through the white room and across the gantry to the baskets, and jump inside. Each basket can hold about three people. They would then smash down on a paddle that cuts a cord and releases the basket they are in. It slides down the cable and away from the gantry, achieving a speed of about 50 mph. A rope net attached to a heavy anchor chain stops the basket from smashing into the ground. In about 30 seconds, the astronauts would be on the ground nearly a quarter of a mile from the launch pad.

After jumping out of the basket, the astronauts would have two choices to continue their escape. They could run into an underground bunker and wait for the emergency to pass, or they could jump in an Army armored personnel carrier and drive away. (NASA teaches all astronauts how to drive this vehicle for just such an emergency.)

## Do astronauts ride the escape baskets in their training?

No. Before every mission, astronauts train with this basket escape technique during a dry countdown. This is usually done about 2 weeks before the actual launch. They get in the shuttle just as if they were going to launch, but there's no fuel in the tanks. The launch control center directs the practice countdown and ends it with a simulated emergency that requires the crew to escape by themselves from the shuttle. They do everything I just described—from unstrapping from their seats to jumping into the armored vehicle—except they don't ride the basket. These are chained to the launch pad so they don't go anywhere after the paddle slap cuts the retaining cord. NASA doesn't want to expose astronauts to the risk of the ride, since the chances are very slight that any astronaut will ever have to use the baskets. (They have never been used for a real emergency.) During their dry countdown training, after they hit the paddle to simulate releasing the basket, astronauts get out of the baskets and take the elevator to the ground. Then they drive to the place where the net stops the baskets and continue the training at that point.

By the way, before the *Challenger* exploded, nobody (astronauts included) had ever ridden the escape baskets. Instead, they were periodically checked by loading them with sandbags. After *Challenger*, NASA demanded that all emergency systems be crew rated through actual usage. This included the baskets. So one astronaut and two rescue personnel volunteered to ride the device. The astronaut was a Marine fighter pilot—Charlie Bolden— which gave rise to an Air Force astronaut's comment, "We used a Marine to verify the baskets were safe for sandbags." There's a lot of good natured, interservice rivalry among the astronauts, and this is an example of the humor that comes from that rivalry.

## Why do they light sparklers at the base of the shuttle when the engines are starting?

Many people think these sparklers are what start the shuttle's liquid-fueled engines. This is not true. The engines have a spark igniter inside their combustion chambers. The sparklers you see swirling around the bottom of the engine nozzles ensure that any hydrogen gas escaping the engines is immediately burned. The worry is, during the engine start sequence, that some unburned hydrogen gas might come out of the nozzles. This gas could mix with the air and form an explosive combination, possibly causing a detonation that could damage the shuttle. The sparklers instantly ignite any gas before it can form a large explosive mixture.

### Why do they start the three liquid-fueled engines before the booster rockets?

A liquid-fueled engine can be turned off. There's no such thing as an off command with a solid-fueled rocket booster. Once it is ignited, it will burn until all of its fuel is exhausted. So, the shuttle's liquid-fueled engines are started early so they can be checked by the computers. If there is a problem with any of them, all of them will be turned off and the mission will be scrubbed. Only if everything is okay will the computers send the ignition command to the SRBs.

The exact timing of the SRB ignition command (about 6.5 seconds after liquid-fueled engine start) allows the stack to spring back vertically from the bending impulse imparted by the three liquid-fueled engines. The force of their thrust causes the stack to momentarily lean away from the vertical (about 18 inches measured at the tip of the external fuel tank). This is referred to as ET twang. When the stack has "twanged" back to vertical, the SRBs ignite and lift-off occurs with the stack in the correct orientation.

As of November 1995, there have been five instances in the shuttle program in which the liquid-fueled engines have not passed their checks and the engines have been shut down. The very first time this occurred was on my first launch attempt (STS-41D, June1984).

### What holds the shuttle to the launch pad when the liquid engines are running?

As we just discussed, the three liquid engines are started about 6.5 seconds before lift-off so they can be checked by computers.

The combined power of these engines is about one million pounds of thrust. That's not enough to lift the shuttle because it weighs 4.5 million pounds, but it's certainly enough to cause it to tip over and explode.

What keeps it on the pad? Each SRB has a skirt with four flanges. On the launch pad, and where each booster stands, are four, large threaded posts that stick upward. These match holes in the flanges around the booster skirt. In prelaunch assembly, a crane lowers the bottom booster segment so the posts slide through the holes. It's just like lining up the threaded posts on a car wheel with the holes in the spare tire. After the booster segment is completely lowered, large nuts are threaded and tightened onto the posts to hold the booster to the pad. Again, this is just like tightening the lugs on a spare tire. It's these eight nuts (four for each booster) that hold the entire shuttle to the pad and keep it from tipping over when the liquid engines are started. Each nut has explosives in it. When the command to ignite the booster is sent, another command is simultaneously sent to fire these explosives. This causes the nut to break in half, and the shuttle then rises into the air.

What do you suppose happens to the used nuts? They've been blown apart and can't be used again, so NASA gives them to astronauts and other personnel as keepsakes of the mission. Each weighs approximately 16 pounds. I use the two halves I was given as bookends.

### Are the SRBs really hollow?

Yes. The center of each booster is a hollow, star-shaped cavity. The propellant is molded in this shape to control the area of the burning surface and thus control the thrust. At lift-off you want maximum power, so you want a large surface area to be on fire. The star shape gives that area. But continuous max power will push the shuttle too fast while it's still in the thick part of the atmosphere. The vehicle could be torn apart by air pressure. So, the SRB power is reduced by a third about 50 seconds into flight. How? As the star burns, it becomes a circle, which has less surface area and therefore produces less thrust.

By the way, the igniter for an SRB is at the very top of the booster, not at the bottom. That igniter is itself a solid rocket. When it's started, it shoots a stream of fire down the center of the main booster, simultaneously igniting the entire surface area of the cavity from top to bottom. This results in an almost instantaneous turn on—from zero thrust to 3.3 million pounds of thrust for each booster in 0.2 second (thats 2 *tenths* of a second). This nearly instantaneous power build-up is so profound astronauts have no doubt that they are leaving the planet when it occurs.

### How much fuel does a shuttle use to reach orbit?

Total solid and liquid fuel consumption from engine start to engine stop is about 3.8 million pounds. During first-stage flight (i.e., while the SRBs are thrusting), fuel consumption averages about 10 *tons* per second. In second-stage flight (i.e., after booster separation), liquid fuel consumption averages about 3,000 pounds per second.

### What does a shuttle launch sound like inside the cockpit?

When the liquid engines start, there's a loud roar and heavy vibrations in the cockpit. I can't think of any normal Earth activity to use as an analogy for these sensations. Some astronauts have compared it with driving a car down a railroad track at 60 mph. Others have likened it to sitting too close to the speakers at a rock concert. Perhaps the best way to appreciate the feeling is to visualize what's happening at engine start. The engines produce a million pounds of thrust, but the rocket is being held to the launch pad by an explosive bolt/nut arrangement. Holding this much power to the earth is obviously going to be loud and teeth-rattling.

When the SRBs ignite, the noise and vibration increase significantly. Nothing I have ever experienced in my aviation career

**Shuttle launch.**

comes close to equaling this noise and shaking. At no time, however, are the vibrations so bad that you can't read instruments, and the noise doesn't prevent you from hearing your fellow astronauts on the intercom. Of course, the helmet and Snoopy cap headphone arrangement insulate astronauts from some of the noise.

From lift-off, the vibrations steadily increase through Mach 1, the speed of sound, where shock waves add to the shaking; the vibrations begin to moderate as the shuttle rises into thinner air. It's when the SRBs burn out and are separated that the most dramatic change in noise and vibration occurs. SRB separation is marked by a bang as the explosive bolts that hold the boosters are fired. This is followed by a flash of fire across the windows as small rockets on the nose and tail of the boosters blow them away from the still-ascending shuttle. At this point the noise and vibration end. The shuttle is about 25 miles high, so the air is very thin, too thin for any wind noise or shock waves to shake you, and, even though the three liquid-fueled engines are still running, there is virtually no noise or vibrations for the rest of the ascent—another 6.5 minutes.

## What does a shuttle launch look like from the cockpit?

Within a few seconds of lift-off, the shuttle performs a roll maneuver so the commander and pilot see a horizon of water and land spinning around their heads. Mission specialists are sitting too far aft to see this. Then, any cloud cover will race into your face and disappear behind you. As the shuttle accelerates, small pieces of the foam covering the ET will be ripped away by the wind blast and will flash by the window. The sky will darken very quickly, becoming totally black by about 2 minutes into flight. That's a strange sight—to see a totally black sky while sunlight is streaming into the cockpit. Around 5 minutes into ascent, the shuttle has pitched (leaned) over enough so the commander and pilot will be able to see the earth. Mission specialist astronauts, however, won't see the planet at all from the forward windows because they are sitting too far back in the cockpit (or, they are sitting downstairs). On my first mission, about 3 minutes into flight, the thought crossed my mind that the rocket could blow up and I could die without ever having seen the Earth from space. So, I twisted my head to look upward through the overhead window. (The shuttle is going into orbit upside down, so the overhead windows look out on the earth.) I got my first, breath-taking glimpse of Earth from space—a royal-blue Atlantic sprinkled with white puffs of clouds.

### What does a shuttle launch feel like?

Besides the increase in noise at lift-off, the crew also gets an instantaneous 1.6-g shove into their seats. For comparison, a typical jet airliner gives you about a 0.33-g shove into your seat during acceleration down the runway for takeoff. In other words, the shuttle's lift-off g's are about 5 times greater than an airline passenger experiences during takeoff roll. The affect of the shuttle's acceleration is mind-boggling. The launch may look slow and stately on TV but in only 4 *seconds*, the shuttle (weighing 4.5 *million* pounds) has already reached 100 mph and in 40 seconds, it's supersonic. It's not your father's Oldsmobile.

The g-forces vary throughout ascent, increasing from 1.6 at lift-off to 2.5 just prior to booster burnout. At SRB burnout (6 million pounds of thrust ends), the g's drop dramatically—to about 0.9 g. From this point on, the ride is glass-smooth and essentially silent. (One astronaut once described this portion of ascent as "an electric ride." I think that's a good analogy. The nearly 1.5 million pounds of thrust from the continued operation of the liquid engines is almost an imperceptible hum.)

As fuel is burned and the vehicle lightens, the g's slowly rise until they reach 3.0 (or about ten times what an airline passenger feels while accelerating down the runways on takeoff) about 7.5 minutes into flight. At this point, the computers begin to re- duce the throttle settings to keep the shuttle from exceeding 3 g. For all the shaking and vibrations that a shuttle has to en- dure, it's really a very fragile vehicle and would tear itself apart if the g's got too much above 3. In fact, there is a shuttle emer- gency procedure for the crew to shut off one of the engines if they don't automatically throttle downward.

Main engine cutoff (MECO) occurs 8.5 minutes after lift-off. In an eye-blink you go from 3 g to weightlessness—and most as- tronauts let out a little cheer. MECO marks the end of the most dangerous part of the mission and that's something to cheer about.

### Do the launch g-forces cause you to black out?

No. In some science fiction movies astronauts are shown black- ing out or having their faces severely deformed by the g-forces of launch. Even at the maximum of 3 g this is not the case on a shuttle launch. The g's are uncomfortable but not disabling or deforming. Basically it feels as if someone is sitting on your chest. Breathing is difficult and speech becomes grunted. Also, it's difficult to twist your head to look at switch panels or reach switches. This is why most astronauts practice raising their arms at various points in ascent to get a sense of the difficulty they would have in fighting the g-forces to complete emergency procedures.

### Do your ears pop when you launch?

No. The reason your ears pop when you fly in an airplane is be- cause air is let out of the plane as it climbs. This is done to keep the air pressure on the inside from pushing too hard on the skin of the plane when it gets up into the really thin air of cruising altitude. (This allows the aircraft manufacturer to make the skin thinner and, therefore, make the plane lighter.) When the plane reaches about 8,000 feet, the release of air is stopped so the passengers can still breath as the plane goes higher. Later, when the plane descends through 8,000 feet to land, air is allowed back inside so the pressure is the same inside and outside. This change in air pressure causes your ears to pop. On a space shuttle, no air is let out as you fly into orbit, or allowed in as you descend to land, so your ears don't pop on launch or landing.

### What would happen if a shuttle went out of control and headed for a city?

This is a very unpleasant thought, but it's certainly something NASA has planned for. The shuttle is equipped with a system known as a *flight termination system* or FTS. Basically, this is a string of dynamite that runs along the side of each solid rocket booster and the external fuel tank. On the ground, a person known as the *range safety officer,* or RSO, watches the launch on a computer screen. Ground radars send information to the RSO's computer so the RSO knows exactly where the shuttle is flying. If something terrible should happen and the shuttle went out of control and headed for land, the RSO would flip two switches to send the detonation signal to the FTS dynamite. The shuttle would be blown up and stopped from threatening any cities. Of course, if the FTS is ever activated, the crew would be killed. While this sounds terrible, you must remember that astronauts have volunteered for the risks of flying a rocket. Civilians sitting in their homes have not. The RSO on duty when the *Challenger* tragedy occurred actually transmitted the FTS destruct commands. Even though the shuttle was right on track and not threatening any cities at the instant it exploded, the two booster rockets, which were still thrusting, were freed by the explosion and did pose a threat. So the RSO sent the destruct commands. In videos of the *Challenger* tragedy, you will see both boosters simultaneously blowing up about 15 seconds after the initial explosion. This was the result of the FTS dynamite detonating. Of course the crew had already been lost, so the FTS detonation had no effect on their survivability.

Some astronauts are vehemently opposed to having an FTS aboard the shuttle. They point out that the chances of a shuttle landing on somebody are small and you might end up killing the crew for no reason. But I disagree. When a shuttle lifts off, it contains about 4 million pounds of very dangerous fuels. If it ever did land on a city, it could kill thousands of people. In fact, the crew families are only about 3 miles away from the launch pad, and there are a couple hundred NASA personnel supporting launch operations who are also close by. So these people are also threatened. In my mind, the only thing worse than a shuttle tragedy that kills the crew is a disaster that kills the crew and the crew's family and many NASA workers, too. It's my opinion that the shuttle needs an FTS to protect people on the ground, be they civilians, NASA workers, or astronaut families. Besides, I believe if a shuttle ever did go out of control, there would be no hope for the crew anyway. It's a fragile vehicle and the stress from an out-of-control maneuver would probably tear it apart.

### Is the shuttle the only rocket that can be intentionally blown up?

No. *All* rockets the United States launches (manned and unmanned) have this dynamite system aboard, but the shuttle design is unique in space history in that the activation of the FTS will kill the crew. Every manned rocket that flew before the shuttle had a similar FTS to protect nearby civilian populations, but the astronauts flying those rockets had a very capable escape system that would probably have saved them if the FTS had been activted. That escape system was in the form of a tower rocket at the tip of their capsules (except *Gemini*, where astronauts had ejection seats). If the RSO had found it necessary to blow up their main rocket (or the main rocket had blown up for any reason), the escape tower rocket would have fired automatically and carried the capsule and its crew away from the explosion. Then, a parachute would automatically have been deployed and the capsule would have parachuted into the ocean. (*Gemini* astronauts would have ejected.) Because it has no similar escape system (see Chapter 7 to find out what the shuttle does have for an escape system), the shuttle became the first manned rocket for which activation of the FTS would mean certain death. When NASA was designing the shuttle, it felt the technology was advanced enough that the shuttle would never be out of control and an FTS activation would never be required. In fact, before the *Challenger* tragedy, many in NASA used this rationale to argue that the FTS should be removed from the shuttle. It was the US Air Force that disagreed with this reasoning and insisted that the shuttle fly with the dynamite. The Air Force demands were rooted in the fact that national policy makes them responsible for protecting civilians from all rocket launches, including NASA's. They didn't share NASA's confidence in the shuttle technology, particularly in the SRB technology, which they felt was most vulnerable to an out-of-control accident scenario. Unfortunately, the *Challenger* tragedy proved them correct.

**Could a landing shuttle be blown up if it went out of control?**

No. The FTS explosives are only on the booster rockets and the external fuel tank. They're not on the orbiter. But people on the ground are no more at risk from a landing shuttle than they are from a regular airplane crashing on them. The chances are very, very small.

**Would the astronauts be able to do anything before the shuttle is blown up?**

There is a light in the cockpit telling a crew that the RSO has sent the first of two commands needed to blow them up. Seeing this light, the crew would immediately attempt to steer the shuttle away from land. But this scenario assumes the crew has control of the vehicle. If they don't have control, then the light is virtually useless. The crew would probably have only a couple seconds to do something, and the only thing they could do would probably kill them anyway. They would press the external tank separation button (ET SEP button) which would blow them off the ET. That would get them away from the dynamite, but in all likelihood, they would immediately spin out of control and crash.

### What are launch windows?

A launch window is the time period during which a shuttle can be launched and still do its mission. Many factors are considered when the launch window is calculated—the need to have daylight at an emergency abort field, the position of the sun when a satellite is being released, the planetary positions if a space probe is being released, and so forth. Usually, rendezvous missions have small launch windows, because there's only a limited opportunity each day when the target satellite will be in the right position to catch.

### How long before a launch does the crew get into the cockpit?

Crews begin strapping into their seats about 2 1/2 hours before the scheduled lift-off. Because the seats are miserably uncomfortable (they're just pieces of flat steel with a thin cushion) and NASA worries that the extreme fatigue of lying in them will affect crew performance, limits for "on the back time" have been established. Theoretically, a crew will never be kept on their backs for more than 5 hours (meaning a launch window will never exceed 2 1/2 hours). I say "theoretically" because most crews would probably consent to wait longer if it appears there's a chance for launch rather than have to try again another day.

### What are launch aborts?

What would happen to a shuttle if, during launch, one of its liquid engines shut down? Depending upon when the failure occurred, it may not be able to reach the planned orbit. In this case, the crew will have to abort the launch. Exactly what happens in an abort is a function of when the engine fails. The crew might return to the Kennedy Space Center—a RTLS abort; land in Africa or Europe—a TAL abort; fly once around the earth and land in the United States—an *abort-once-around* (AOA); or fly into a lower but safe orbit—an *abort-to-orbit* (ATO).

As you might imagine, any abort is a very serious situation and any mistake in an abort could kill the crew. For this reason NASA's training program devotes hundreds of hours to teaching astronauts how to fly aborts. This is why, after successfully reaching orbit, astronauts joke among themselves that they've wasted hundreds of hours in launch emergency simulations. Nobody complains, though. We would just as soon see *all* emergency training as wasted time.

## What is an RTLS abort?

If an engine failure occurs very early in ascent—within approximately 2 minutes of lift-off—the crew will be unable to make it into space or even to an emergency field in Africa or Europe. They will have to select RTLS abort, and fly back to the Kennedy Space Center runway. This will be a very unusual (and terrifying) flight. Imagine you are an astronaut, and just as the SRBs separate (approximately 2 minutes into flight), your center liquid-fueled engine shuts down. You dial RTLS on the abort switch and press the abort button. At this moment, you are upside down (head toward the earth) traveling toward Africa at several times the speed of sound. The shuttle immediately flips itself to point its engines in the direction of travel and the nose toward Florida. Now, your head is toward the sky and the external fuel tank is underneath you. But you are traveling backward over the ocean. Remember, you were going thousands of miles per hour to the east when the abort was selected. Even though the shuttle nose is now pointed toward Florida, it'll take several minutes to cancel that velocity. During these minutes, the shuttle will slow down until it's at zero speed about 42 miles above the Atlantic Ocean. At this point it's a 1-million-pound brick falling straight down at hundreds of feet per second. Then, it'll slowly begin accelerating to the west—toward Florida. When the fuel tank is empty, it's jettisoned and the shuttle starts a glide to the runway.

Besides an engine failure, there are other emergencies that may require an RTLS, a cabin pressure leak, for example. You wouldn't want to continue flight into orbit when the air you need to breathe is leaking out.

### What is a TAL abort?

About 2 minutes, 15 seconds into ascent, the shuttle will reach a point where an engine failure would still prevent it from reaching orbit but now it would be high enough and fast enough to make an emergency landing in Europe or Africa. This is called a TAL abort. Basically, this is a complete mission contained in a 35-minute flight. When the crew selects TAL on the abort button, the shuttle rolls to put the fuel tank toward the earth. But it continues to fly straight ahead. When the navigation system determines the shuttle is within gliding range of the landing site, the engines are shut down, the tank is jettisoned, and the orbiter reenters the atmosphere and glides to a landing. Imagine crossing the Atlantic Ocean in just 35 minutes. That's what a shuttle will do during a TAL abort. (It takes the supersonic Concorde about 3 1/2 hours to make the same flight.)

### What is an AOA abort?

About 5 minutes into flight, an engine failure would still prevent a shuttle from reaching a stable orbit, but it will be going so fast and be so high it could make one loop around the earth. This is an AOA. Again, this is like a complete mission now compressed into about 90 minutes. Landing could be made at several places: Edwards Air Force Base in California; White Sands, New Mexico; or back at the Kennedy Space Center.

### What is an ATO abort?

About 5 minutes, 30 seconds into ascent, the ATO abort window opens. At this point, the shuttle will be going so fast and be so high that an engine failure won't prevent it from reaching a safe orbit. The crew will fly an ATO. Their final orbit might not be the planned orbit, but at least they would be in a temporarily safe orbit—above 120 miles. Later, they may be able to use the OMS engines to boost them closer to their intended orbit.

## Has any mission ever had an inflight abort?

An ATO is the only abort a shuttle crew has ever had to perform on an actual launch. On mission STS-51F, at 5 minutes and 46 seconds into launch, the center engine failed. Fortunately, this was late enough in the flight that the crew was able to abort nearly into their planned orbit. This was a day NASA and the crew were very lucky. As it turned out, there was nothing really wrong with the engine. It shut down because of some faulty temperature sensors. If the engine had failed a little earlier, the crew would have been faced with a TAL, which would have been significantly more dangerous. (Anytime you're landing a 100-ton glider at an unfamiliar field, the risks are much greater.)

## Are there aborts for two engine failures?

If two liquid-fueled engines fail (depending upon when they fail), the crew might still be able to do an abort with the remaining single engine. Mission Control will keep them informed of single-engine abort options. In many cases, though, two engine failures will result in the loss of the shuttle. It won't be able to reach a runway and will crash into the ocean. Theoretically, in this situation the crew could bail out.

## Are there aborts for SRB failures?

If something went wrong with a SRB, the crew would attempt a *fast sep* (meaning a fast separation) from the external tank. There's a button on the console between the commander and the pilot that fires explosives that release the tank from the belly of the shuttle. Since the SRBs are attached to the fuel tank, getting away from it also gets you away from the boosters. Theoretically, the crew could then glide the shuttle to a landing or to a position where bailout would be possible. In reality, however, a fast sep would probably kill you. Most engineers think the orbiter would be immediately flipped out of control and would crash. Also, SRBs have a nasty habit of catastrophically failing within *seconds* of the onset of a problem, meaning the crew would have no time to react to the failure and even try a fast sep. In every sense of the words, you are along for the ride while the SRBs are burning. If something goes seriously wrong with an SRB, the crew and shuttle would almost certainly be lost.

All of the aborts that are selectable on the abort switch are referred to as intact aborts, meaning the orbiter should be able to land intact with an intact crew. A fast sep is not certified as an intact abort. In fact, as I mentioned earlier, it's a last chance effort that would probably result in death. In a blatant misuse of the English language, NASA has defined a fast sep and other similar aborts as contingency aborts. In moments of dark humor, astronauts joke that contingency abort procedures are just something to read while you're dying.

### What is the meaning of the tech-talk during launch?

Now that you understand what launch aborts are, you can make some sense out of all the technical gibberish you hear during a shuttle launch. Many of these calls refer to aborts. I've assumed the shuttle in question is *Atlantis*.

"Houston, *Atlantis*, roll program." The commander makes this call about 10 seconds after lift off. He's telling Houston the shuttle is rolling to the correct launch azimuth, the direction to fly the mission orbit. Really, Mission Control can see the roll on their displays. They don't need the crew to tell them it's happening. But the call is planned anyway, because astronauts want to check their radios as soon as possible to make sure they are still working after all the vibrations of lift-off. Saying "Roll program" for a radio check is as good as saying anything else. The Capcom, the astronaut in Mission Control who talks to the crew, answers, "Roger, roll." At all points in every shuttle mission, the crew and Mission Control repeat what was just radioed between them. This is done to ensure that nobody has misunderstood what was said.

"*Atlantis*—go at throttle up." Mission Control makes this call about a minute into flight to tell the crew that everything locks good after their passage through the thick part of the atmosphere. About 30 seconds earlier, the autopilot had reduced the power of the three liquid-fueled engines so the shuttle wouldn't be damaged by going too fast through the lower atmosphere. "Throttle up" means the engines have returned to full power.

"*Atlantis*, Houston, performance . . . nominal." The other possibility with this call is, "performance . . . low." Mission Control is telling the astronauts how the SRBs are performing. There are a lot of variables in how these things will perform; for example, the air temperature and exact mix of factory ingredients affect the thrust. You might get a set of hot boosters that give you more thrust than was expected or a set of cold boosters that are a little weak. Mission Control has the tracking data to see what kind of power the boosters are producing—nominal or low—and can inform the crew. The crew wants to know this information because it will affect some emergency procedures.

"*Atlantis*, Houston, two-engine Moron [pronounced more-own]." This is a Mission Control call that the TAL window (Trans-Atlantic abort window) has opened. In other words, the shuttle is high enough and going fast enough so that if one of its engines failed, it would be able to fly across the Atlantic Ocean on its two functioning engines and make an emergency landing at the planned TAL airfield. Moron is an air base in Spain, but the call could be to other abort sites in Africa or Europe.

"*Atlantis*, Houston, negative return." This is an announcement the RTLS abort window, which opened at lift-off, has now closed. In other words, the shuttle is now so far from Florida and going so fast to the east, it cannot turn around and fly back to the Kennedy Space Center runway. From this point on, if there is an engine failure, the shuttle must fly one of the straight-ahead aborts: TAL, AOA, or ATO.

"*Atlantis*, Houston, press to ATO." Mission Control is informing the crew that the ATO window has opened.

"*Atlantis*, Houston, single engine Moron." Now, the shuttle could lose two engines and still make a TAL abort on the last remaining engine. It's around this point in an ascent that the crew begins to breathe a little easier. They rightfully think it would have to be a *really* bad day to lose two engines and now, even if they did, the shuttle could continue to fly on the last engine to a safe landing at the TAL field.

"*Atlantis*, Houston, press to MECO." MECO means main engine cut-off and is an abbreviated way for Mission Control to tell the crew they are now going fast enough and are high enough that if an engine failed, they wouldn't have to select any aborts. This is a call that really gladdens astronauts' hearts. Most of the danger of launch is now behind them.

### Why does the shuttle go into orbit upside down?

This is done so the commander and the pilot will see the earth's horizon as early as possible. During launch, the upside-down shuttle is continually pitching (tilting) over so as to ultimately be nearly parallel with the earth's surface. This pitching maneuver means the earth's horizon will slowly come into the astronaut's view from above. Having such a view would help the crew to manually steer the shuttle into orbit in the very unlikely event that the automatic guidance system failed. In such an emergency, they would point the shuttle's nose slightly above the horizon and hope they achieved a safe orbit (good luck!). If the shuttle went into orbit right-side-up, this procedure wouldn't work, because the crew's view would be blocked by the big belly tank, which extends too far forward for the commander to be able to look down and see the horizon.

### Does a shuttle launch damage the ozone layer?

The exhaust gases of the SRBs chemically react with the earth's ozone layer, resulting in some damage to it. But the damage is minuscule compared with the damage done by other human activities. In fact, scientists estimate that eight shuttle launches a year contribute less than 3 *hundredths* of a percent of the total human-caused ozone depletion per year.

### Who flies the shuttle into orbit?

Nobody. The shuttle flies itself into orbit. Its computers are loaded with all the information it needs for the autopilot to steer it into orbit. The crew can take over from the autopilot at any time and fly the shuttle with the control stick, but if they ever had to do this, it would be a very serious emergency. The shuttle has to fly a very precise path to get into orbit. Think of it as an airliner taking off from Los Angles for Hawaii, with only a couple *seconds* of extra fuel. If you were on that airplane, would you rather have the pilot flying by the seat of her pants or an autopilot in control that can precisely determine position, winds, optimum fuel consumption, and all the other things needed to get to Honolulu's airport? Clearly, you would have more faith in the autopilot. It's the same on a shuttle.

On a nominal ascent, the crew will make only one switch change and even that has nothing to do with how the shuttle will fly. The switch merely changes the way the attitude indicator (the ball at the center of the instrument panel) displays the shuttle's orientation, making it easier for the crew to interpret their attitude. Otherwise, they will merely be watching the ascent and be ready to intervene if there are emergencies.

Anyone interested in a home computer software program that approximately simulates the cockpit displays seen during a shuttle launch should contact NASA engineer Dan Adamo at adamod@iapc.net.

### What is Launch Control?

Launch Control is located at the Kennedy Space Center and is the NASA team that directs the shuttle's 3-day countdown to launch. At tower clear (i.e., just a few seconds into launch),

Launch Control will pass control of the ascent to Mission Control in Houston, Texas.

## What is Mission Control?

Mission Control is located at the Johnson Space Center in southeast Houston and controls the mission from tower clear to wheel stop (at landing). At that time, the Kennedy Space Center resumes control in powering down the orbiter, making its hazardous systems safe, and returning it to Kennedy Space Center (if it has landed elsewhere).

Actually the term *control* in Mission and Launch Control is a slightly misleading one. If everything is nominal, nobody controls the shuttle during launch or reentry. It uses its own computers to fly into and out of orbit. Launch and Mission Control, as well as the astronaut crew, watch their displays and will manually intervene only to correct malfunctions or update navigation parameters.

## If shuttles are launched from Florida, why did NASA build Mission Control and the astronaut training facilities in Texas?

Even very young children ask me this question, proving that fourth graders are smarter than politicians. Why wouldn't NASA want all of its shuttle operations located at the place where the rockets are assembled and launched? In a word, the answer to this question is *politics*. Johnson Space Center isn't named *Johnson* by accident. In the early days of the manned space program, then Vice President Johnson was a huge NASA supporter and wanted his home state to have a lion's share of the glory of the space program—and its money—so he mandated that Houston would be the home of NASA Mission Control. Never mind that such a dispersement of operations merely makes a complex and expensive business even more complex and expensive. The majority of astronauts silently curse the bad luck that made Johnson a Texan and hot, humid Houston our home. Why, we lament, couldn't he have been from someplace with great weather and great scenery, like Colorado or Utah or Arizona or New Mexico?

## Where does space begin?

There's no abrupt change in the earth's atmosphere that marks the transition to space. As you go higher and higher, the air gets thinner and thinner. At 5 miles altitude, atmospheric pressure has already dropped to less than half of sea level pressure, and by 50 miles altitude it's down to about one thousandth of sea level pressure. For all intents and purposes, when you're above 50 miles, you're in space. That doesn't mean there's *nothing* above 50 miles. There are still trillions of molecules of various gases. It's just that they are spread so thin that the environment at these heights is essentially a vacuum.

## When is the shuttle actually in orbit?

Technically, the shuttle has achieved orbit when the three liquid-fueled engines shut down—about 8 1/2 minutes after lift-off. When you consider it has consumed about 4 million pounds of solid and liquid fuels in this time (an average of over 7,000 pounds per second), you can appreciate what astronauts know— a shuttle launch is a controlled explosion. But the orbit a shuttle has achieved after the main engines stop is an ellipse with a high point (apogee) of 185 miles (varies with the mission) and a low point (perigee) of about 40 miles. This perigee is too low for the vehicle to stay in orbit. Atmospheric friction will pull it down. So, to achieve a *sustainable* orbit, the crew must do another burn about 45 minutes after lift-off. This burn is accomplished with the twin OMS engines and increases the shuttle speed about 150 mph, which is enough to raise the perigee so that a circular orbit of around 200 miles is attained. (The shuttle launches with about 19,000 pounds of OMS fuel inside the tail assembly to use for this orbit insertion burn, as well as for other orbit adjustments and for the deorbit burn.)

**What happens to the shuttle's empty fuel tank?**

After main engine shut down, the fuel tank is jettisoned. Since it's going the identical speed as the shuttle, it's in orbit, too. But remember, that initial orbit isn't sustainable. As the tank flies toward perigee (the orbit low point), most of its 66,000 pounds will be burned up. The pieces that survive will impact in a remote area of the Pacific Ocean.

**Who has the toughest job during a shuttle launch?**

It's not who you might think—somebody on the astronaut crew. In many serious, time-critical emergencies, the astronauts don't have enough data to make a proper decision. It's the Mission Control flight director (FD) who, most likely, will be faced with life-and-death decisions. The FD is the person who supervises the entire Mission Control team. He will take the recommendations of that team and make his decision. Any delay in a decision, or an erroneous decision, could easily result in the deaths of the astronauts. As you might imagine, every astronaut has great trust in the flight director and the rest of the Mission Control team. If you didn't, you would never be able to climb aboard a shuttle.

**Where are the astronaut families during launch?**

Even before the *Challenger* explosion, NASA had an emergency action plan to care for astronaut families in the event of a tragedy. One of the high priorities of the plan was to ensure that the press had no access to grieving families until they wanted to grant such access. Unfortunately, prior to the *Challenger* tragedy, this disaster plan had some flawed procedures that led to Christa McAuliffe's family being at a site close to the press. As a result, they were photographed in their moment of greatest grief, while watching the explosion that killed their daughter. Afterward, NASA corrected the plan to ensure adequate care and press isolation are achieved for astronaut families in any future disasters.

Under this plan, the immediate family members—spouses and children—watch the launch from the roof of the Launch Control Center, about 3.5 miles from the launch pad. The extended families—mothers, fathers, siblings, aunts, uncles, and other relatives—watch from a separate viewing area. At both sites there are astronaut escorts who are ready to care for the families if tragedy strikes. The press is located at a site completely removed from either of these family viewing areas.

# CHAPTER 3

# Space Shuttle Orbit Operations

### If weightlessness is a free fall, do orbiting astronauts feel like they are falling?

I can't speak for all astronauts, but occasionally, when I was in the lower cockpit, I would experience a sensation of falling—like the floor was a sheer cliff and if I didn't hang on, I would fall off of it. The sensation was rare, but whenever it occurred, it was very powerful, so powerful in fact that it was difficult to resist reaching out to grab something as an anchor. Since I never had a similar sensation in the upper cockpit, I assume the illusion was caused by a lack of visual perspective (there are no windows in the lower cockpit). Other astronauts have reported periodic, short-lived sensations that they were tumbling.

### What's it like to float in the shuttle?

No doubt about it, weightlessness is fun—excluding the vomiting and backache (see Chapter 5, Space Physiology). With just a touch of a finger you can send your body flying across the cockpit, and heavy items (e.g., the shuttle's 97-pound removable seats), that were a struggle to deal with in Earth simulations can be effortlessly moved.

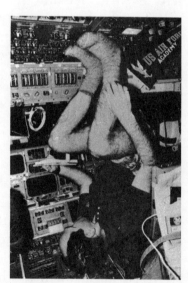

*A weightless astronaut.*

It does take a little time to get used to weightlessness, though. For the first hour or so of my rookie mission, I found myself constantly fighting my body orientation. I wanted my head toward the cockpit ceiling and my feet on the floor. After a while, the brain finally begins to accept the fact there is no up or down in orbit and you get used to being in all attitudes.

## Would it be possible to hurt yourself when flying across the cockpit?

Yes. If you pushed too hard from one wall you could break something on impact with the opposite wall. This isn't likely to happen, however. Astronauts rapidly learn to control their body movement with their fingers.

## Could an astronaut get stuck floating in the center of the cockpit?

Theoretically, if someone could hold you in the center of the mid-deck cockpit and release you with absolutely zero residual force and there was no air circulation to move you, and the shuttle had absolutely zero forces on it, then yes, you could be stranded. But none of these conditions exists on the shuttle. Air is blowing around and the forces on the shuttle are never truly zero. Even in free-drift (i.e., no thrusters firing), the shuttle is still being hit by a very, very thin wind of molecules, which will exert a force of micro-g's (millionths of a g) and ruin the pure weightlessness needed to keep something stranded.

## Would it be possible to kick a switch into the wrong position while floating around the cockpit?

Kicking a switch to the wrong position on a space shuttle could ruin your whole day, but it's not likely to happen. For one thing, astronauts are very careful about the position of their hands and feet. For another, all of the switches are set between rounded hoops, or wickets. These extend further from the panels than the switch levers themselves and therefore protect the switches. You must reach between the wickets to change the switch. Finally, for very critical switches, other protection is added. For example, a cover must be raised to get to the switch, or the switch requires an outward pull before it can be moved.

## Have you seen any UFOs?

This is the second most-asked question I ever hear. ("How do you go to the bathroom in space" is the most asked question). But, no,

I've never seen anything in space (or on Earth) that I thought was an alien spaceship, and, frankly, I don't believe anybody else has ever really seen, talked to, or been kidnapped by alien creatures. Having said this, I do believe there's intelligent life elsewhere in the universe. Scientists estimate there are 200 billion billion stars in the known universe. That's 200,000,000,000,000,000,000,000 stars. Even if you assume only one out of a *quadrillion* of these have planets with intelligent life, you still end up with 200,000 alien home planets. So, it's easy for me to believe that we are not alone. But I believe if and when that life ever visits earth, it will make a *significant* contact with humans. It won't be like the news stories in which somebody out in the middle of the desert sees a flying saucer but nobody else does. Why, I ask myself, would an alien civilization go to the trouble of building a spaceship and flying around the universe looking for life so it can hide from it? Imagine how laughable "Star Trek" would be if its mission was to "boldly go where no person has gone before, to seek out new life and new civilizations . . . and then hide from them."

Though I have had people tell me they have watched TV stories about astronauts who have seen UFOs, I have never heard a first-hand account by *any* astronaut of such experiences. I suspect these astronaut TV accounts are actually based upon astronaut statements taken out of context. Many astronauts on many missions see unusual things—strange vertical flashes (rare vertical lightning discharges), distant flashing lights (tumbling pieces of space junk), colorful streaks of light in the upper atmosphere (the possible exhaust plume of a ground-launched rocket), moving formations of lights (specks of ice and frozen urine floating near the shuttle). In the hands of a slick TV producer these reports of colors, flashes, streaks, and formations can suddenly turn into accounts of alien encounters. Until an astronaut personally tells me he has seen something he believed was extraterrestrial in origin, I will remain very skeptical of these alleged astronaut sightings. I would encourage anybody who sees a TV or press report that names an astronaut as having seen an extraterrestrial craft to record the astronaut's name and send it to the E-mail address mentioned in the preface to this book. I will do my best to track down the astronaut and do a personal interview for publication in a future revision of this work.

### Are astronauts given training on how to deal with any aliens they might encounter?

No, and I can't imagine how anybody could design a training program for such an encounter. If any aliens had ever knocked on the shuttle hatch while I was in space, I would have grabbed the microphone and called, "Houston, we have a problem."

### Can astronauts see other satellites from the shuttle windows?

Yes. Most satellites have very shiny surfaces of metal foil or solar cells. These reflect the sunlight, and if the shuttle is in the right position (so that the satellite is being seen against the blackness of space), astronauts will be able to see this reflection. On my second shuttle mission, a satellite we released looked like a bright star, even when we were a couple hundred miles away. Remember, the light is reflected sunlight. Unless they are targets for rendezvous and/or retrieval, satellites don't have real lights on them.

### Can you see the hole in the ozone from space?

No. Ozone is invisible to the human eye. Scientists can see it only with special instruments. But you can see the atmosphere whenever you look at the horizon, night or day. During the

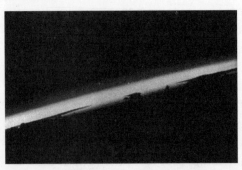

day, it appears as thin layers of various shades of blue. At night these layers appear as different shades of gray. Whenever the atmosphere is seen, though, it's a sobering sight. All Earth life is sustained by it, yet it's incredibly thin.

*A sunset highlights the earth's atmosphere.*

## Can you talk in space?

Sound will not travel in a vacuum, so it's impossible to talk in space (even if you could somehow stay alive to try). The only way for a space-walking astronaut to communicate is via radio.

The silence of space is something most science fiction film writers choose to ignore. For example, in most films, outside explosions are invariably heard by the crew. In reality, an atomic weapon detonated near a space shuttle (or a photon torpedo detonated outside the starship *Enterprise*) would never be heard.

But what about talking *inside* the space shuttle? This is no different than talking on Earth, because the air in the shuttle is just like it is on Earth and sound travels the same. The shuttle atmosphere is about 80% nitrogen and 20% oxygen at a sea level pressure of 14.7 pounds per square inch (psi). Astronauts who are together in the upper cockpit or in the mid-deck area can talk to each other just as people in a room on Earth would be able to talk to each other. To talk between the upper cockpit and the mid-deck requires astronauts to use the intercom or to raise their voices to be heard over the noise of the cabin fans (which circulate air to cool the cockpit electronics).

## Do the shuttle engines continue to run in orbit?

No. Once orbit velocity is achieved, the engines are no longer needed. The laws of nature—orbit mechanics—keep the shuttle from falling back to Earth.

## What does it sound like in an orbiting space shuttle?

Orbit flight is very quiet. This is most noticeable during the sleep periods when mission control has signed off and the crew is asleep. The only sound is the soft whooshing noise of the cabin cooling fans. It's a remarkable quiet, because it flies in the face of all terrestrial experience. On Earth there is always sound associated with speed, and, usually, more speed brings more noise. But in orbit, you are traveling at nearly 5 miles per second in velocity and there's no noise at all—no engine noise, not even a whisper of wind noise. There were times, as I was looking from the windows, when this extreme quiet overwhelmed me with a powerful illusion—that I wasn't moving at all, that I was hovering and it was the earth that was slowly turning underneath me.

### Can you sneeze in weightlessness?

Yes, and you do. Immediately after MECO (the first time you experience weightlessness), dust, lint, and other tiny items that had been hidden in various cracks and crevices will immediately float into the air and cause some astronauts to sneeze. Even during the mission, after the air filters have had time to filter out some of these initial irritants, there is always some debris (tiny bits of food, human hair, lint, etc.) floating around that will occasionally provoke a sneeze.

### Can you blow your nose in space?

Astronauts who are inside the space shuttle use regular tissues for blowing their noses. If you are in a spacesuit, however, there is no way to do this. Nose blowing is done to clear congestion. It's not done to stop a nose from dripping or running in space. In weightlessness, nothing drips or runs.

### What does water look like in weightlessness?

All fluids, water included, form a perfect ball when they are free to float in the cockpit. Why? Because the molecules making up the water are all pulling on each other. This molecular attraction—called surface tension—produces a sphere. Surface tension is present in fluids on earth, too, but the force of gravity is so vastly greater that it overwhelms this weak molecular force.

It's fun to watch surface tension at work in weightlessness. If you slowly squeeze a stream of water from a drink container, the stream quickly coalesces into a perfect ball. (Fluid leaving your body in the form of urine exhibits the same behavior.)

Space shuttle astronauts have to be careful when playing with water. If the water ever got sucked into our air conditioning system it would be sprayed onto our electronic equipment and could cause an electrical short circuit.

### How would a candle burn in weightlessness?

To better understand the answer to this question, let's review how a candle burns on Earth.

The wax is the fuel, and the air supplies the oxygen for combustion. As the air above the candle gets warm, it gets lighter and rises. This causes more air with more oxygen to be pulled in at the bottom of the candle. The candle continues to burn because the flame is constantly being supplied with fuel (wax) and oxygen (air).

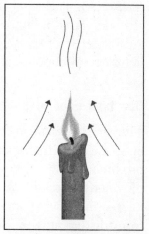

*On Earth, a candle heats the air, making it lighter (convection). Oxygen is continually supplied to the flame.*

Now, let's examine what would happen in a space shuttle. First of all, the flame will not be tear-shaped. It will be a ball. Why? Because in weightlessness there is no convection. The only reason hot air rises on Earth (convection) is because it's lighter than the surrounding air. The hot molecules move further apart, the air becomes less dense, and therefore, it rises. But can anything be lighter than anything else in weightlessness? No. Everything, hot and cold air included, weighs the same—zero. This means the candle will consume the oxygen that's next to the wick and then go out. In other words, you would see a brief ball of fire. (This explanation assumes there are no air currents to stir the air and provide more oxygen to the wick. Aboard the shuttle, fans blow air around, so the candle would probably continue to burn.)

*In weightlessness, there is no convection. Heated air will not rise. The candle goes out.*

## How fast does a space shuttle travel?

A shuttle orbits at approximately 17,300 mph. That's nearly 5 miles per *second*, or 10 times faster than a rifle bullet. At this speed, it takes a shuttle about 10 minutes to travel from Los Angeles to New York. While these are impressive numbers, astronauts don't really have a sense of their tremendous speed. They're too high in altitude. Any military astronaut will tell you there is a much greater sense of speed when going 600 mph at tree-top level in a fighter than clipping along at 17,300 mph in a shuttle orbit.

## How high does a space shuttle orbit?

The shuttle orbit altitude varies with the mission. The highest apogee (orbit high point) flown as of November 1995 was approximately 385 miles, during the Hubble Space Telescope deployment mission. Scientists wanted it as high as possible to keep atmospheric drag to a minimum and to keep gas molecules from distorting the view. The lowest perigee (orbit low point) was approximately 100 miles, during a rendezvous mission with the *Mir* space station. The low perigee was temporarily needed to bring the shuttle into the optimum position to catch the *Mir*. Most shuttle orbits are circular around 185 statute miles.

## Can the shuttle change its orbit, climbing and diving like a space fighter in sci-fi movies?

No. Its ability to change its orbit (altitude and inclination) is severely limited by the physics of spaceflight. Look at this simple drawing:

**Solid Rocket Booster and Liquid Main Engines**

**OMS Engines**

*Comparison of the speed potential of shuttle rocket systems.*

The long arrow, or vector, represents the direction and speed of an orbiting shuttle, while the tiny arrow (drawn in scale to the big arrow) represents an orbiting shuttle's ability to *change* its orbit with its onboard OMS maneuvering system. In other words, if a shuttle burned all of its excess OMS fuel through its twin 6,000-pound thrust engines, it could change its speed about 270 mph, or the graphic equivalent of the little arrow, and that's all. These two arrows tell us the shuttle orbit is basically fixed at MECO, that is, when the main engines shut down (that's the 17,150-mph big vector). With that tiny 270-mph speed reserve, the OMS system—the little vector—can raise or lower the shuttle's orbit about 150 miles or change its orbit inclination (tilt to the equator) by about 1 degree. That's a far cry from the maneuverability of Luke Skywalker's X-wing fighter.

## Would a shuttle stay in orbit forever?

No. Even though we refer to space as a vacuum, there are still trillions of molecules of various gases where a shuttle orbits. It's just that they are spread exceedingly thin. Still, impact with these particles causes aerodynamic drag, which slows the shuttle and results in orbit decay (a slow drop in orbit altitude). Unless it is reboosted, every satellite will eventually succumb to atmospheric drag and fall to Earth. The estimated orbit life of a space shuttle at its typical orbit altitude of 185 statute miles is approximately 1 month. That's no big deal because a shuttle will have to come back to Earth after a week or two anyway. For other satellites that are supposed to stay in space for a very long time—like the Hubble Space Telescope and the future space station—drag is a very big deal and provisions are made to periodically reboost these spacecraft. The Hubble Space Telescope is reboosted by the shuttle during its maintenance visits. The future space station will have its own jets for periodic reboosts. Fuel for these jets will be brought from Earth by unmanned Russian tanker rockets.

## Is it true the only human-made object you can see from space with the naked eye is the Great Wall of China?

This is a misconception that has been around since the very beginning of the space program. In fact, on a popular TV game show, this was given as the correct answer to one of their Space category questions. But it's the wrong answer!

It's true that astronauts can see the Great Wall of China with the unaided eye, but that's not the *only* human-made object they can see. For example, they can also see cities, fires, very large ships, air and water pollution, vapor trails of jet planes, airport runways, and very large buildings. Long, straight roads

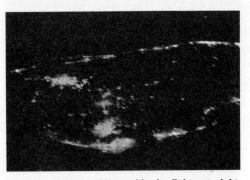

Florida peninsula ablaze with city lights at night. The view is from the west at the bottom to the east at the top, with Tampa and Saint Petersburg visible in the foreground and Orlando in the background.

through deserts, jungles, and snow are also easily seen. But nighttime city lights are the easiest and most spectacular human-made objects to view. They look like lava flows, because the city centers are bright and the lights of the suburbs trail away from them.

Most people are surprised to hear that astronauts can see so much from orbit. They think the shuttle is orbiting at an extreme distance from Earth. Actually, shuttle astronauts are very close to the earth. Depending upon the mission, a shuttle is somewhere between about 100 and 400 miles above the earth. That sounds like a great distance but think of it this way: When you are in an airliner at about 35,000 feet altitude, you can see to a distance of about 200 miles. That's about the distance a shuttle astro-

200 miles

200 miles

35,0000 ft

The vertical distance of a typical shuttle orbit is about the same as the horizontal line of sight from an airliner.

naut is from the earth. If you can see something on the horizon from the airplane window (e.g., city lights), then astronauts—the same distance straight up—are also going to see the object.

### Can you see the borders of countries from space?

The Mexico-U.S. border is visible in different land-use patterns south of the Salton Sea. The view is toward the south.

Much has been made of the fact that astronauts see the world free of humankind's petty territorial divisions. Actually, the borders of some countries are visible from a shuttle orbit because the people on either side of the border use the land differently. For example, one side might be irrigated and have farm fields, while the other side is undeveloped. It will appear that there is a line drawn between the countries. That's the border. Portions of the Mexican–US border in southern California look this way. Different farming techniques on either side make the border visible from space.

### Does the moon look bigger when seen from a space shuttle?

Moonrise from shuttle orbit.

No. It doesn't look any different than it does when you watch it from Earth. Remember, the moon is 240,000 miles from Earth. A shuttle in a 240-mile orbit has only traveled approximately one *thousandth* the distance to the moon. If you go one thousandth of the distance closer to anything, does it look any bigger? No.

You can easily demonstrate this by measuring the distance across a room and dividing that distance by 1,000. Let's say the distance across your living room is 20 feet. Stand on one side of the room and pick an object on the other side. Now, step one thousandth of the distance toward that object (about a quarter of an inch). Does the object look any bigger? Of course not. And neither will the moon from a space shuttle window.

### What do the planets look like from the space shuttle?

They, too, don't look any different than when you see them from earth. They appear as bright stars. Even at their closest approach, our planetary neighbors, Venus and Mars, still remain approximately 25 and 50 million miles away, respectively If you go a couple hundred miles closer to something that's 25 or 50 million miles away, does it look any different? Clearly not. So astronauts see the planets exactly as you see them from Earth.

### What do the stars look like from the space shuttle?

After reading the answers to the last two questions, I think you can now answer this one. The stars are *trillions* of miles away, so obviously they don't look any bigger from the window of a shuttle that's only gone a few hundred miles closer. They do look different, though, in that they don't twinkle. Because the light isn't being distorted by our murky atmosphere, they appear as fixed dots of light. Also, since astronauts are far above any light pollution, they see many more stars than the typical city-dwelling Earthling. From space, the sky is *frosted* with stars. The Milky Way looks exactly like its name: milky.

Another thing about the stars that you can see better from space is their color. Most people think that all stars are white, but there are orange, red, and bluish-white stars, too. Some of these can be seen from Earth (e.g., Aldebaran— the eye of the bull in the constellation Taurus—is an orange star). But star colors are better seen from space.

**Is there any way on Earth to see the stars as astronauts see them?**

Actually, there is. Climb a tall mountain that's away from any city lights and you'll see a sky very similar to what astronauts see. I have done this many times, while hiking in the mountains of the West. From the top of a 12,000 foot New Mexican mountain, the night sky is ablaze with stars. They still twinkle, but you see a frosting of them, and the Milky Way looks like spilled milk. Pilots in high-flying aircraft also get a view of a night sky similar to what astronauts see.

**Can you see stars in the daytime when you're in space?**

Only one star—Sirius(the brightest star)—is visible from the day side of an orbit. The sun's reflection off the clouds, ocean, and the shuttle itself is so brilliant the pupil of the eye contracts, making it impossible to see dimmer star light. The bright planets—Venus, Jupiter, and Saturn—are also visible during the day.

This is another one of those space realities that most science fiction movie directors ignore. These movies typically show stars in the windows, even though there is a lot of ambient light. In reality, the only way the Star Trek *Enterprise* crew could see stars from their windows would be to turn off the cockpit lights when they are in a planet's shadow or in deep space.

**What does a black hole look like from a space shuttle?**

First of all, nobody positively knows whether black holes exist. Theoretically, under some conditions, a star could collapse inward and become so dense that even light could not escape its intense gravity field. If this happened, it would be a black hole. If black holes do exist, nobody, including shuttle astronauts, would be able to see them because light cannot radiate from them. Someday scientists might be able to positively confirm the existence of black holes by observing their gravitational effects on surrounding objects that are visible through telescopes. It's certain though that no black holes are near us, because their strong gravity would affect our entire solar system.

## What does the sun look like from the space shuttle?

Just like on Earth, you can't look directly into the sun from a shuttle without damaging your eyes. But in your side vision, you can tell the sun is pure white. There's no yellowish color as is typically the case when it is viewed from Earth.

To keep astronauts from being injured by ultraviolet (UV) light, all but one of the shuttle windows have UV filters built into the glass. The only window that doesn't have this protection is the small, circular window at the center of the side entry hatch. This window was intentionally made optically pure to allow all wavelengths of light to pass through it in case a cockpit experiment required a UV view of something. A portable UV filter can be Velcroed over this window so astronauts will not be harmed if sunlight should shine through it.

## What does Earth look like from the space shuttle?

To appreciate an astronaut's view, the first thing you need to understand is that shuttle astronauts are very close to Earth—too close to see the Earth as a globe. Those famous photos you've seen of Earth hanging in space like a blue-white marble were taken by *Apollo* astronauts who were tens of thousands of miles away. For shuttle astronauts, the earth *fills* the windows. It's hugely close, and they only see a fraction of its surface. Actually, they can see about 1,200 miles in all directions. That's a lot, but it's not enough to see an 8,000-mile-diameter Earth as a ball. A good way to appreciate exactly how close to Earth a shuttle orbits, is to scale an orbit around a familiar spherical object. For example,

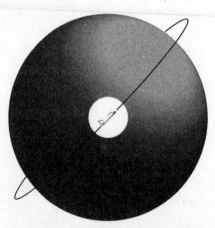

*At any one time, a shuttle astronaut sees only about 2.5 percent of the earth's surface. The circle represents the limit of an astronaut's line of sight.*

if we used a 12-inch globe to represent an 8,000-mile-diameter Earth, then a 200-mile-high orbit *in scale* would only be about 0.3 inch above the globe's surface. That's 3 *tenths* of an inch. Another example is to use a large orange—about 3 1/2 inches in diameter—to represent the earth. In this case, a scaled shuttle orbit would be a little less than 0.1 inch, or about the thickness of the skin of the orange.

Because of its vast oceans and cloud cover, the earth's predominant colors are blue and white. For the brief moments you are over land, color is a function of the geography. Deserts are brightly visible in earth tones of brown, tan, red, and orange. Jungle areas appear very dark, almost dark bluish in color. On missions that fly over the Arctic and Antarctic areas, the color is obviously white because of the snow and ice.

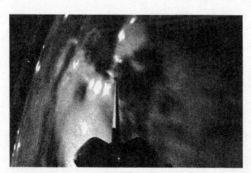

Lightning seen from the space shuttle. The white areas are lightning, and the vertical structure is the shuttle's tail.

### Can you see lightning from space?

Yes, but only at night. You don't see forks of lightning, but the flashes illuminate the clouds, making white explosions. Over areas of intense thunderstorm activity, it's possible to see hundreds of flashes a minute.

In the accompanying photograph, taken from a shuttle's payload bay window, the white blobs are the lightning flashes. (They look lightly smeared because the camera was hand-held and the photo was made at a slow shutter speed.) The tail of the shuttle is the pointed object in the foreground.

### Can shuttle astronauts see entire continents at one time?

Astronauts on one shuttle mission flying at 380 miles altitude over the midwestern United States were able to simultaneously see the lights of Los Angeles, Seattle, New York City, and Miami, so it is possible to see across the North American continent at one time. However, an astronaut's perspective isn't the same as viewing a classroom globe, on which the continents appear to be islands dwarfed by vast oceans on all sides.

**Can you see shooting stars from space?**

Yes, but, again, they are only visible at night. Shooting stars are meteors that are being incinerated by the heat of air friction at an altitude of 30 to 50 miles, so astronauts always look *down* to see them. Because they have such a wide, clear view, astronauts see many more shooting stars than do Earth-based observers.

**Can you see meteors floating in space?**

As a child, I saw science fiction movies in which astronauts could see meteors tumbling by their spaceship. In the real world, the speeds involved make such viewing impossible. Remember that the shuttle is traveling at 10 times the speed of a rifle bullet. A meteor might be moving at a speed of 10 times the shuttle speed, or 100 times faster than a bullet. If you can't see a bullet, you're obviously not going to see a meteor whizzing past.

*An aurora seen from the space shuttle.*

**Can you see the aurora borealis, the northern lights, from space?**

On high-inclination missions (missions in which the orbit is tilted steeply toward the earth's poles), the northern (and southern) lights make a spectacular display. They appear as fuzzy, changing curtains of green-yellow and reddish light. The actual phenomenon is occurring around 50 miles altitude, so astronauts look down to see the show.

### What causes the auroras?

The northern and southern lights are caused by the sun, but they are not the result of sunshine (in fact, you can't see them during the day). During solar flares, electrically charged particles are thrown into space. Some of these get trapped in the earth's magnetic field and are pulled into the atmosphere where that field enters the earth—near the north and south poles. An electrical reaction in the upper atmosphere causes these charged particles to emit light.

### Can you see a rainbow or fireworks from space?

No. These events are too small to be seen from a shuttle.

### Can you see a tornado from space?

*Ocean thunderstorms seen from the shuttle*

You can see the giant anvil heads of thunderstorms that spawn tornadoes, but you can't see the actual funnel cloud.

### Can you see hurricanes and typhoons from space?

Yes, very easily. They appear as giant swirls of clouds funneling into the eye of the storm.

*A hurricane photographed from the shuttle*

### Can you see volcanoes erupting from space?

Yes. The smoke and ash from volcanoes stretches for hundreds of miles across the sky, making it very easy to see the volcano itself.

*An astronaut's view of an erupting volcano*

### Could astronauts see the oil fires from the Persian Gulf war?

Yes. They could see the fires at night and the smoke during the day.

### What's the most beautiful thing you can see from space?

In my opinion, sunrises win first place among all the incredible spectacles an astronaut witnesses. Just imagine being in the earth's shadow and looking out the window and seeing nothing but blackness. Then, about a minute before the sun rises, an eyelash-thin arc of deep, deep purple appears in the east. This is caused by the atmosphere. It acts as a prism and splits the pure white of sunlight into the colors of the rainbow and indigo is the first color to outline the eastern horizon. As the sun rises higher (but it's still below the horizon), various shades of blue appear below the indigo. Tints of orange follow the blue and those, in turn, are followed by bands of red, until a wide arc of the eastern horizon is defined by this magnificent color bow. It reaches peak beauty just seconds before the sun actually comes up. When that happens, the color is instantly washed away and the earth becomes visible beneath you.

Sunsets produce a similar show, except in reverse. The wonderful part of spaceflight is that you get to witness this spectacle approximately every 45 minutes. One sunrise and one sunset divide the shuttle's 90-minute orbits.

## What's the weirdest thing you've ever seen from space?

On my last mission, we were in a relatively low orbit—about 150 statute miles. At night, when we looked from the rear windows into the payload bay, it appeared that the shuttle had been transformed into a ghost ship. A foot deep fog of glowing white light covered every surface that was pointed into our orbit path. If we hadn't been prepared for it, it would have been a disconcerting sight.

What we were seeing is a phenomenon known as atomic oxygen glow. In the upper fringes of the earth's atmosphere exist atoms of oxygen (not molecules as we breath on Earth). As these strike an orbiting spacecraft, they react with the surface material and produce a glow.

## Is it easy to know where you are when you look at the earth from the shuttle?

No. Except when you are flying over obvious landmarks (e.g., the Florida peninsula, Baja, the boot of Italy, etc.), it's very difficult to know exactly where you are. Remember, most of the time you're over the ocean, which makes it impossible to know your position on the earth. We carry a lap-top computer with orbit traces similar to Mission Control's map, as well as geographic maps, so we can find ourselves.

## Do astronauts have time to look out the window?

Yes. How much window time will be a function of the mission. If the primary objective of the flight is to release a satellite, then the crew will probably have a lot of time afterwards to watch the sights. On the other hand, on a *Spacelab* mission, in which the crew is in a payload bay module conducting experiments, the astronauts might be so busy they won't have significant window time. Is astronaut sightseeing a superfluous waste of money? Absolutely not. Oceanographers, meteorologists, and geologists are constantly hounding NASA for more astronaut photography.

They have learned that no robotic, earth-watching satellite can compare to an astronaut's ability to observe and record unusual ocean, weather, and geologic events. On a typical mission, a crew will return with thousands of photos from their window time—a bonanza for earth scientists.

### Why does the shuttle usually orbit upside down?

Imagine what a shuttle has to endure when it's flying around in space. Where the sun is shining on it, it'll be over 250 degrees above zero, and where there are shadows, it'll be about 200 degrees below zero. This extreme temperature difference can cause expansion and contraction and bending in the skin of the shuttle. All of this stress can be minimized by flying the shuttle with its top pointed toward the earth. So, when mission objectives don't require otherwise, the shuttle will orbit in this optimum attitude for thermal conditioning. It's also a great attitude for astronauts to watch the earth, because there are two windows on top of the shuttle cockpit. Note that I said the shuttle flies upside down when mission objectives don't require otherwise. It can orbit in *any* attitude to support *any* mission experiment, but every several hours the crew may have to interrupt data collection to put the shuttle in a more benign attitude for thermal conditioning purposes.

### Does everybody on a crew sleep at the same time, or is somebody always on watch?

This varies with the mission. On some missions, such as *Spacelab* missions, there are usually two shifts of astronaut workers, so somebody will always be awake. On other missions, such as a mission to deploy a satellite, everybody will work and sleep at the same time.

Regardless of whether the crew is awake or asleep, Mission Control will be carefully monitoring the shuttle systems. Also, if all the crew is asleep and a critical failure did occur, an alarm will sound to wake them.

## Who steers the shuttle when the crew is awake? When they are asleep?

Nobody. Once the shuttle is in orbit, the laws of physics keep it in orbit. No engines have to be on and nobody has to steer it around the earth. It's held in orbit by gravity and as long as gravity doesn't fail (not very likely), it'll stay in a fall around the earth. So if everybody aboard the shuttle were asleep and everybody in Mission Control went out for pizza, the shuttle would continue to circle. As for altitude control, the crew usually sleeps with the autopilot engaged, and it will command the appropriate reaction control system (RCS) jet firings to keep the shuttle in the preferred orientation.

## Are the astronauts always awakened with music by Mission Control?

Wake-up music is always transmitted at the end of a crew sleep period. The Capcom selects the song to be broadcast. Selections have run the gamut of music types: classical, new age, country and western, rock, and so on. It would be unusual, though, for a crew to be awakened by the music. In most cases the crew is awake before the end of the sleep period. When you're living in a small space with five or six other people, it's difficult to sleep soundly. As soon as one person is up, everybody is awake, and that usually happens before the wake-up music is sent. Also, the speakers on the shuttle are so small and so low-fidelity, the crew sometimes won't even recognize the tune.

## Do astronauts worry about getting hit by meteors?

No. The risk from meteors is very small. Over nearly 40 years, Earthlings have launched thousands of satellites and space probes and none is known to have ever suffered a significant meteor impact (all satellites are hit by tiny pieces of space dust). So, compared to the immediate hazards they face (like riding a machine propelled by 4 million pounds of highly explosive chemicals), the risk of a meteor impact is a distant worry to astronauts. NASA does take steps, however, to minimize even this small risk. They have delayed a shuttle mission during a known period of increased meteor activity (i.e., during a meteor shower). Such showers occur when the earth's orbit crosses the path of a comet.

## Do astronauts worry about hitting other satellites?

Scientists estimate that there are approximately 4.4 million pounds of human-made orbiting objects, including approximately 3,000 burned-out booster rockets, within about 1,200 miles of the earth's surface. Air Force radars keep track of about 8,000 pieces of this material. But there are also thousands of other pieces too small to be detected by radar. All of this material is traveling in many different orbits at various altitudes. The low-orbiting pieces are traveling at about 17,300 mph, which means a head-on collision could occur at a combined speed of nearly 35,000 mph. Clearly, if such a collision ever occurred with a shuttle or a spacewalker, and depending upon the size of the object, death could be the result. Do astronauts worry about this? Not really. You have to remember these objects are swirling around in a *huge* sky. The chances of a collision are very, very small. A few steps of simple arithmetic will suggest how small. Let's assume there are 10,000 objects that could threaten a shuttle and, instead of being at *random* altitudes, they are at the *same* 200-mile altitude of an orbiting space shuttle. What is the average distribution of these objects?

$$\text{The surface area of a sphere} = 4\pi^2$$

We know pi is 3.14. So, the only other thing we need to plug into this formula is the radius of the sphere. For our question, that's the radius of the earth plus 200 miles.

$$\text{Radius} = \text{Earth radius} + 200 \text{ miles} = 3,950 \text{ miles}$$
$$+ 200 \text{ miles} = 4,150 \text{ miles}$$

Therefore the area of an imaginary surface at 200 miles altitude is 216,314,600 square miles. For simplicity, let's round this off to

200 million square miles. This means there are 10,000 objects randomly crossing paths on an area equal to about 200 million square miles—a distribution of about one object in every 20,000 square miles. This means, in all likelihood, you won't even *see* another object, much less run into one.

Another way of thinking about this is to calculate, on average, how many objects would be in an area the size of your state. I live in New Mexico, which has a surface area of 121,000 square miles. If you think of that as a square, the square would measure about 350 miles on each edge. At an average distribution of one object per 20,000 square miles, this means that at any one instant, there would be about six satellites occupying an area equal to the entire state. Think of these objects as six cars randomly driving around an asphalt lot the size of the state of New Mexico and you're in one of those cars. What are the chances you would even *see* another car, much less get hit by one? Obviously, they are very small. This is why astronauts don't spend a lot of time worrying about running into other satellites. And, remember, for these computations, I assumed the objects were *all* at a shuttle's exact altitude. Actually, they will be at different altitudes. In other words, they will be flying over and under the shuttle. So, the chances of getting hit are even smaller than these calculations suggest.

Another point to consider is the amount of time an astronaut is exposed to the risk of a collision. For a space shuttle astronaut, it's very brief. The shuttle can stay up only about 2 weeks. But what about a space station that will be in orbit for decades? Obviously the chances of its being hit are much greater. Still, no matter how you look at it, the chances of being hit by another object are so small, it's not a big worry for NASA or astronauts.

### Does NASA do anything to minimize the chances of being hit by other satellites?

Just because the chances are small doesn't mean a collision can't happen, and NASA does take precautions to minimize the possibility of a life-threatening impact. Specifically, they minimize the time when the cockpit windows are into the velocity vector (when they are facing the direction the shuttle is traveling). While there's no guarantee a shuttle couldn't be hit from behind by an object in a crossing orbit, the chances are greater that a damaging impact would come from the front. Also, on some missions, Mission Control has directed the crew to partially close one of the payload bay doors to protect the radiator surface that's on the inside of the door. (There's one radiator system on the inside of each door.) These radiators circulate the liquid freon used to cool the shuttle's electronics, and if an object ever penetrated both radiators, the crew would be in a serious emergency, because the electronics will eventually overheat and fail.

### Has any shuttle ever been hit by a piece of space junk?

The 7th shuttle flight landed with a very small pit in one of the windows. NASA analyzed this tiny mark and determined it was probably made by a paint chip that had flaked off an old rocket.

### Does NASA watch the sky to see if any satellites might threaten the shuttle?

NASA doesn't, but the Air Force does and will warn NASA if a close approach is likely to happen. In fact, sometimes NASA will delay a launch by a minute or two if Air Force computers show that an on-time launch will put the shuttle close to a known object. Also, after the shuttle is in orbit, the Air Force continually predicts how close the shuttle will pass to other objects and tells NASA if there is any possibility of a collision. Normally, these warnings are given hours in advance, so there's plenty of time to react. NASA merely directs the crew to make a very small thruster firing—maybe only a couple feet per second in velocity. Over hours, that's enough to cause the shuttle to miss the other object by many miles.

**Does the shuttle have a radar to warn the crew they are approaching another satellite?**

No. The shuttle has a radar used in rendezvous operations, but that device has such a narrow field of view it would be useless as a tool to search the skies for approaching junk.

**What happens if an astronaut gets seriously sick in space?**

It's difficult to imagine anyone could get seriously sick in orbit, given the number of medical checks that are done prior to launch. But doctors can't see everything, and it's certainly possible an astronaut could suffer some life-threatening illness like appendicitis, heart attack, bleeding ulcer, and so on. Also, there is the potential for space accidents such as electrocution, the bends (a very real threat with spacewalks), and toxic gas inhalation. So, what would happen if an astronaut did come down with a life-threatening illness or suffer a life-threatening accident? NASA would make every effort to get the shuttle safely out of orbit and the patient to adequate care. The key, here, is *time*. The shuttle can't just fly out of orbit anytime somebody is sick. Depending upon its orbit inclination, there are only three to six deorbit opportunities each day that will bring the shuttle to a landing in the United States. Of course, a decision could be made to deorbit to *any* city where adequate health care is available, in which case there would be more deorbit opportunities. This would be a tough decision, though. It could be very dangerous to try to land the shuttle at an unfamiliar field. Many fields, particularly in foreign countries, do not have the electronic landing aids that help astronauts make precision approaches, and their runways may be shorter than desired. Obviously NASA wouldn't want to put the lives of the other crewmembers at significant risk trying to save one person.

**Does the shuttle have any medical equipment to treat injuries and illness?**

Yes. There is an emergency medical kit that contains an assortment of pain-killing, anti-nausea, and other drugs as well as hypodermic needles, bandages, an IV bag, and basic diagnostic equipment.

### How can you give an IV in space?

Since the IV fluid is weightless, there's no way it can drip into a person's body. Instead, it's done by squeezing the bag with a blood pressure cuff. The squeeze forces the fluid into the vein.

### Is there a doctor aboard every shuttle flight?

No. In fact, there were no doctors aboard any of my three missions. But, since approximately 20% of mission specialist astronauts are medical doctors, many missions will have one.

### On missions with no doctor, is anybody trained in medical care?

Yes. Because it is much more likely a crewmember might face a non-life-threatening medical problem like space sickness, a tooth ache, the flu, and so on, even someone with minimal medical training might be able to render assistance. (NASA isn't going to end a billion-dollar space mission to bring somebody back to see a dentist.) So, as part of the training for any doctor-less crew, two crewmembers are given a little medical training (e.g., taught to give an injection by practicing on oranges). They also have the help of a shuttle checklist that explains some basic medical procedures, and advice from the Mission Control flight surgeon is just a radio call away. Personally, though, I think I could endure any amount of pain before I would permit a Marine pilot, whose training consisted of having once injected a piece of fruit, to come at me with needle in one hand and a checklist in another, saying, "Bend over."

### Can you give CPR in space?

Yes, but with considerable difficulty. Remember Newton's law: For every action there is an equal and opposite reaction. As soon as you press down on somebody's chest, you'll float away. To prevent this, NASA has designed a harness that will attach both the patient and the person giving the CPR to the wall. The final arrangement has the doctor straddling the patient. The harnesses hold each in place so the doctor can push down on the patient's chest without floating away. It is highly unlikely that anybody facing an immediate need for CPR (e.g., a heart attack or electrocution victim) could ever be saved by such a procedure (it's rarely successful even on earth). By the time somebody could locate the harness among the lockers, get it on the patient and get the patient clipped to a wall, and get dressed in her own harness and clip it to the wall, the patient would probably be dead. But I'm certain that if the circumstances warranted CPR, the crew would try to administer it.

### Has anybody ever died in space?

No American astronauts have ever died in space. Three *Apollo* astronauts died in a launch pad fire in 1967, and seven shuttle astronauts were killed in 1986 in the *Challenger* explosion but neither of these tragedies occurred in space. (*Challenger* exploded while still in the earth's atmosphere at approximately 46,000 feet.) The only humans that have died in space were three Soviet cosmonauts who were killed in 1971. Shortly after their deorbit burn—and while still in space—their capsule experienced a cabin pressure loss. They were not wearing pressure suits and died of the effects of the cockpit decompression.

### What would astronauts do if somebody died in orbit?

NASA has no checklist to address the disposition of an astronaut's body, so my answer to this question is conjecture. If someone died inside the shuttle, I'm certain the body would be returned to Earth. The surviving crewmembers would probably stow the deceased in the airlock and would deorbit at the first available US landing opportunity. If someone died outside the shuttle (a more likely event because of the inherent dangers of spacewalking), I suspect his body would be retrieved by the other spacewalker and returned to Earth.

### Has anybody ever been buried in space?

No. Even the three Soviet cosmonauts who died in space in 1971 were buried on Earth. Their capsule continued its reentry under automatic control and landed normally. The bodies of the cosmonauts were recovered from the capsule and later buried.

While nobody has ever been buried in space, the partial ashes of Star Trek originator Gene Roddenberry were carried aboard shuttle flight STS-52. A NASA official thought Roddenberry's influence on America's space program through the Star Trek series had been so profound he should be afforded a space flight—even if in death. NASA arranged with his family that the shuttle would carry the ashes. The remains were not released into space, however, but returned to Earth and given back to Roddenberry's family.

### Do astronauts carry suicide pills in case they get stuck in orbit?

Absolutely not! To hand out a suicide pill would be like saying, "I don't know if this rocket is going to work." We believe that the NASA team has given us a spaceship that will bring us back alive and we intend to come home.

### Do astronauts carry any weapons aboard the shuttle?

No. Many science fiction movies depict astronauts carrying weapons, usually as a defense against aliens, but none are carried aboard the shuttle. We don't anticipate being boarded by aliens.

### Would a gun fire in the vacuum of space?

Yes. The primer (the cap the hammer strikes) does not require oxygen to function and neither does the powder in the bullet cartridge. However, if an unrestrained spacewalker ever did fire a pistol, the recoil would send her tumbling head-over-heels backward.

## How can somebody go outside for a spacewalk without all the air in the shuttle being sucked into space?

Spacewalks are done via an airlock. The space-walking astronauts enter the airlock from the shuttle's downstairs cockpit. After they're dressed in their space suits, the inner hatch is closed, isolating them from the rest of the orbiter. Now they are free to depressurize the airlock and exit through the payload bay hatch for their spacewalk. When that's complete, they reverse the process. They float into the airlock from the payload bay, close the outer hatch behind them, repressurize the airlock, rejoin the rest of the crew via the inner hatch, then get out of their suits.

## Can anybody do a spacewalk at anytime?

No. On a typical mission, only two crewmembers have been trained for a spacewalk, and their two spacesuits will be the only ones aboard. Also, the preparation protocol for a spacewalk takes several hours. For example, to minimize the potential for the bends, the spacewalkers must breath pure oxygen for a minimum of 40 minutes prior to depressurizing the airlock.

## Why couldn't space-walking astronauts have rescued the Italian tethered satellite that broke free of the shuttle?

On a mission in 1996, a shuttle carried an Italian satellite that was reeled out of the payload bay on a 12-mile-long cord. The cord was made of a material that is six times stronger than steel, but nevertheless, it broke during the test. Many people wondered why the astronauts didn't do a spacewalk to retrieve this very expensive satellite. The reason it was abandoned in orbit was that the astronauts had no procedures or tools to make such a rescue. Without those, any retrieval attempt would have been extremely dangerous. If the cord got wrapped around a space-

walker or looped around the shuttle, it might have been disastrous. For example, if it got tangled around the payload bay
doors and they couldn't be closed, the astronauts would have
been unable to reenter the atmosphere and would have died in
orbit. If the satellite had been in a higher orbit, NASA might
have trained a crew and later sent another shuttle up to rescue
the satellite. However the craft was in such a low orbit that it
reentered the atmosphere within a month and burned up. A
postflight inspection revealed that the break was caused by an
electrical short circuit that generated enough heat to weaken
the tether.

### How do you keep time in space?

On Earth, we conduct our lives by reference to Earth local time
(e.g., Pacific, Mountain, Central, Eastern). On the shuttle, using
local Earth time would be chaos. Astronauts pass into a new
time zone about every 4 minutes. Even tying the flight plan to
the fixed reference of Greenwich Mean Time (GMT) wouldn't
work. For example, suppose the shuttle mission was to release a
satellite 8 hours, 13 minutes, and 16 seconds after lift-off.
Launch is scheduled at midnight, or 00:00 GMT. A flight plan
tagged to GMT would then specify satellite release at 08:13:16
GMT. This sounds like a perfectly workable arrangement—until
a rain cloud moves over the pad and delays the launch by 12
minutes and 32 seconds. Now the flight plan is wrong. Satellite
release is no longer at 08:13:16 GMT. It's at 8:25:48 GMT.
Everything on the flight plan would have to be adjusted by the
launch delay, creating a paperwork nightmare.

   To avoid these problems, NASA uses the only time reference
that's not going to be changed by delays—*Mission Elapsed Time*,
or MET. At the instant of launch, a computer clock starts counting upward from 00:00:00:00. This is MET, wherein the digit
groups represent days, hours, minutes, and seconds. Every 24
hours, the day digit of the clock is incremented by one, and the
seconds, minutes, and hours begin another count upward from

zero for the next MET day. So, by referring to the MET clock, the crew can figure out where in the flight plan they are and what needs to be done. In the example given, the satellite release would *always* be at 00:08:13:16 MET, regardless of launch delays.

### How does a spacesuit work?

A spacesuit gives an astronaut the same protection provided to all Earthlings by our atmosphere: oxygen, carbon dioxide removal, radiation protection, temperature control, and pressurization. High-pressure tanks contain oxygen to breath. Carbon dioxide (a potentially deadly by-product of respiration) is absorbed by canisters of lithium hydroxide, as are used in submarines. Radiation protection is provided by the suit's covering. Temperature control is achieved by an article worn next to the skin, called a *liquid cooling and vent garment* (LCVG). This has tiny plastic tubes woven into the fabric. A suit pump circulates water through these tubes to pick up body heat. The water is then cooled in a sublimator. Also, the LCVG has air tubes woven into its arms and legs to circulate oxygen around the body. Because shuttle spacewalkers have complained of their hands getting too cold, NASA is experimenting with electrically heated spacesuit gloves.

The one suit function you might question is *pressurization*. It's not apparent to us Earthlings, but we have a significant weight of air on our bodies—almost 15 pounds per square inch at sea level. As that pressure is reduced, the gas in our body tissue and blood (primarily nitrogen) will come out of

**A spacewalking astronaut.**

solution. At very low pressure (e.g., in the vacuum of space), our blood will boil and death will occur. (This is very similar to the decompression sickness that divers can suffer if they ascend too fast.) The spacesuit keeps enough air pressure (4.3 pounds per square inch [psi]) on an astronaut's body to keep this from happening.

An everyday analogy to this blood boiling phenomenon is the effect on a bottled soft drink when the cap is removed. Bubbles suddenly appear in the fluid. It was the pressure on the fluid—held by the bottle cap—that had kept the gas in solution. The atmosphere is our bottle cap. It holds gases in solution in our blood and tissue. In space, where there is no atmosphere, the spacesuit has to serve at this bottle cap.

### Can spacewalking astronauts eat and drink?

Yes. An energy bar and a water bag are Velcroed inside the suit just below the neck. By ducking his head downward, a spacewalker can take a bite out of the food bar and suck some water through a straw.

### Can you scratch yourself on a spacewalk?

You can't use your hands to scratch yourself anymore than you could scratch yourself in a suit of armor. A spacesuit is like an inflated tire. It holds a force of 4.3 psi. That may not sound like much, but it's enough to rigidize the stiff material that makes up a spacesuit to the consistency of steel. Of course, a spacewalker might be able to scratch by rubbing the itch against the inside of the suit. The nose can be scratched by ducking the head down and rubbing against a Valsalva device that's on the inside of the helmet. (A Valsalva device enables an astronaut to plug her nose so she can pop her ears by blowing.)

### How much does a spacesuit weigh on Earth?

The entire spacesuit weighs about 275 Earth pounds.

**How long will a spacesuit keep an astronaut alive?**

The life of the suit's consumables (oxygen, carbon dioxide absorbent, water, and battery power) is a function of the size of the astronaut and how vigorously he is working. It is estimated that an average size astronaut engaged in moderate labor would exhaust the consumables in about 9 hours.

**If a space-walking astronaut died inside a spacesuit, would the body decompose?**

As the deceased whirled around the earth, the external suit temperature would vary from minus 200° F in the earth's shadow to plus 250° F in sunlight. How this extreme temperature cycle outside would affect a lifeless body inside is unknown, but the deceased would probably freeze solid. If so, the body would not decompose.

**Is the spacesuit self-contained, or do astronauts rely on oxygen through an umbilical cord?**

The spacesuit is completely self-contained. The oxygen is carried in a pressurized tank in the backpack portion of the suit.

**Why is space black?**

The only reason Earthlings see a blue sky is because the molecules in our atmosphere scatter the shorter wavelengths of the sun's light (the blue wavelengths) more effectively than other wavelengths. In space, there are essentially no molecules to cause such scattering, and therefore, space has no color.

**How cold or hot is it in space?**

When we talk about temperature on earth, we are generally referring to the air temperature. Because there is no air in space, there can be no similar temperature measurement. Of course, the radiant energy of the sun shining on something affects the temperature of the object, so sunshine on something in space will heat its surface to approximately 250° F (depending on its surface reflectivity and composition). In a shadow, the surface temperature of an object will be approximately minus 200° F.

### How long can a shuttle remain in orbit?

As of November 1995, the longest mission a shuttle has ever flown has been 16 days. The limiting factor on shuttle flight time is electrical supply. The shuttle has no solar power or batteries. Its sole source of power is fuel cells. These convert liquid oxygen and liquid hydrogen into electricity (and water). A shuttle must be back on Earth before it runs out of these consumables, which means it has a maximum flight time of about 2 weeks (and even then, such a long mission would require an extra set of oxygen and hydrogen tanks).

### Can you talk to your family while you're in orbit?

On missions that are more than 10 days in duration, each crewmember is allowed one 15-minute radio conference with his or her family. The conference is private.

### Does the shuttle have a fax machine?

Yes, but it can only receive faxes. Astronauts can't send hardcopy faxes to Earth. However, astronauts can send and receive E-mail via a lap-top computer that can transfer data into and out of the shuttle's communication system. (Sorry, but those of you with E-mail capability in your home can't address mail to an orbiting astronaut. Right now, the link is only with Mission Control, but I suspect in the not too distant future NASA will have an orbiting shuttle E-mail address.)

**Do you take your own cameras in space?**

No. The only equipment ever used on a shuttle is NASA equipment, cameras included.

**What personal items can an astronaut carry into space?**

NASA's rules are very strict on this subject ever since some *Apollo* astronauts carried stamps to the moon and later capitalized on their rarity. Now, astronauts can carry only 20 personal and 10 official items. Personal items are things like religious articles, wedding rings, and Uncle Harry's lucky rabbit's foot. Official items are for organizations. Some examples are school flags, museum articles, and hometown memorabilia. In either category it is forbidden to carry stamps or coins or other collectibles whose value will be enhanced by virtue of a ride aboard a taxpayer's rocket. All items must be approved by NASA headquarters. Then, about a month before launch, they are packed by NASA in a shuttle locker. Astronauts are forbidden to carry anything personal in their pockets. Back in the old days, astronauts who smuggled golf balls to the moon and corned beef sandwiches into orbit were folk heroes of sorts. Now, that type of behavior is frowned upon. I doubt if you smuggled some of grandma's cookies aboard you'd get in trouble, but try taking a stack of stamped envelopes to "cancel" in orbit for sale to collectors, and you can kiss your career goodbye. In fact, you may find yourself getting fitted for a prison uniform.

**Can a shuttle get stuck in orbit?**

There is a *very* low probability this could ever happen. The primary deorbit system is the two OMS engines. If one of these failed, the other would still be sufficient to deorbit the shuttle. If *both* OMS engines malfunctioned, there are four aft-pointing *reaction control system* (RCS) jets that could be fired to slow a shuttle out of orbit. If all of these failed, it might even be possible to turn the shuttle around and use the three forward pointing RCS jets to make the deorbit burn.

Other system failures—besides propulsion system failures—could also theoretically trap a crew in orbit. For example, you can't reenter the atmosphere if the payload bay doors aren't closed. But on these critical systems there are redundant electrical power sources and redundant electric motors. Also, the crew is trained to do spacewalks to manually winch the payload bay doors closed. With all of these system backups, it's very difficult to imagine that a shuttle could get stuck in orbit.

### If a shuttle *is* stranded in orbit, could anything be done to save the crew?

Probably, yes, but the chances of rescue are going to depend upon how close to launch another shuttle is and the stranded crew's remaining oxygen and carbon dioxide absorbent. Assuming there is at least a week of these consumables remaining, and there's a shuttle on the pad, I believe NASA would be able to pull off another *Apollo 13* type rescue and rush another shuttle into orbit to save the astronauts.

*An artist's conception of a rescue sphere in use.*

### How could astronauts be moved from a stranded shuttle to a rescue craft?

Just reaching the stranded shuttle isn't going to mean a rescue unless you can get the stranded crewmembers across the vacuum of space to the functioning shuttle. In this situation, the rescuers would carry *personal rescue spheres* to make the transfer. A rescue sphere is a windowless, pressurized, oxygen-filled ball about a yard in diameter. Two spacewalkers would carry these plastic and fabric cocoons to the crippled ship's airlock. There, the stranded astronauts would curl up and be zippered inside. The spheres would then be pressurized and carried to the rescue craft. You might wonder why the rescuers wouldn't just carry spacesuits for everybody. Storage limitations prevent this. The suits are very bulky, while the deflated rescue spheres are hardly bigger than a garment bag.

### Could a Russian rocket rescue astronauts aboard a stranded shuttle?

Probably not. The Russian *Soyuz* capsules only have room for three crewmembers. At least one of these seats would have to be occupied by a cosmonaut pilot, meaning there could be a maximum of only two return seats. Any stranded shuttle would have a minimum of five people aboard.

### What speed would you have to go to leave orbit and never come back to Earth?

Increasing an object's speed to 25,000 mph will cause it to permanently escape the earth's gravity. This means a shuttle would have to add about 8,000 mph velocity to its orbit speed to escape. On a typical mission, its OMS system is capable of producing an extra 300 mph in velocity, nowhere near that needed for gravity escape.

### What do astronauts miss most in space?

A shower! After a lifetime of taking a daily shower or bath, going a week or two without one feels disgusting. About the third day in orbit, astronauts would kill for a hot shower.

### Can the astronauts talk to Mission Control at any time?

Even when the shuttle is on the opposite side of the earth from Houston, the crew remains in contact through a network of geosynchronous communication satellites called *tracking, data, and relay satellites* (TDRS). There are small gaps in the TDRS coverage

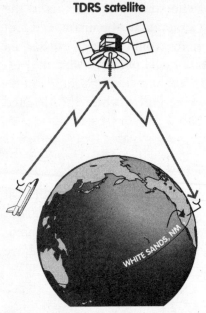

**TDRS satellite**

*Shuttle signals reach the ground via TDRS Geosync Satellites.*

and communication can sometimes be interrupted when the shuttle is in certain attitudes, but, for the most part, astronauts are in continuous contact with Houston. When a crewmember presses the microphone button and calls Mission Control, the signal doesn't go directly to the ground. It is transmitted to whatever TDRS is in view (there are three). This satellite then relays it to the TDRS system ground station at White Sands, New Mexico. Here it is electronically processed and retransmitted to another geosync satellite (different from the TDRS) which, in turn, relays it to Mission Control in Houston, Texas.

Even at the speed of light (about 186,000 miles per second), the distance and processing time involved in this relay are great enough to cause a several-second delay in the communication. In fact, by the time shuttle data appears on Mission Control's computer screens it's about 6 seconds old. Mission Control's transmissions to the crew follow the same communication path in reverse.

### Is there a lock on the shuttle hatch?

Yes. But it's not to keep out any aliens. After achieving orbit, a shuttle crew will install a device on the hatch handle to prevent it from being *inadvertently* moved from the inside. The crew commander also has the option of carrying a padlock to prevent *intentional* movements of the handle. This latter lock would guard against the extremely remote possibility that a suicidal or sleepwalking (sleep-floating) crewmember might try to intentionally open the hatch.

The hatch handle only needs to be turned through one revolution for the hatch to open outward. Obviously, if this ever occurred in space, it would be instant death for the crew. With a sea level pressure of 14.7 psi inside the cockpit and a vacuum outside, there would be *tons* of force ready to violently throw open the hatch as soon as it was unlocked. The crew would be sucked into space and killed instantly.

Locking the hatch handle with a padlock so it can't be turned is a prudent safety procedure. The hatch has no function in space. Astronauts use a different hatch (the airlock) to leave the shuttle for spacewalks. So, nothing is jeopardized by the lock once the shuttle is in orbit. The lock is removed for launch and landing because, during these mission phases, a crew may have to do an emergency evacuation through the hatch and you wouldn't want to be fumbling with a padlock.

The option of the padlock does raise this question: "Are there suicidal astronauts who might open the hatch in space if it wasn't locked?" Who knows? NASA doctors conduct a psychiatric evaluation on all astronauts and shuttle passengers, but how much does that reveal? On at least one occasion, a non-NASA scientist astronaut became very depressed when his experiment did not function properly. Could he or someone else become so depressed they would be suicidal? The chances are probably very, very small, but why take the chance?

But what about the airlock hatch? It's not padlocked. Couldn't someone suicidal open it with the same affect as opening the side hatch? No. The airlock hatch opens *inward*. Just as there are tons of force trying to push open the side hatch, there are tons of force trying to keep the airlock hatch closed. It would be impossible for anybody to open it, except in the course of a normal spacewalk.

## Why did NASA adopt the very dangerous design of an outward opening side hatch?

It would be much safer for orbit operations if the shuttle side hatch was designed to open inward. Then, cabin pressure would deliver tons of force to keep it closed. The reason it uses the more dangerous outward opening design is rooted in history. In 1967 three astronauts were killed in a launch pad test when their *Apollo I* capsule caught fire. The accident investigation determined that the astronauts' escape was fatally delayed by a complicated hatch-opening mechanism. NASA wanted to ensure the shuttle had a simple, outward-opening hatch to facilitate emergency ground escape.

### Does the shuttle have a tool box?

Originally, there was some debate among NASA engineers as to whether various repair tools should be flown aboard the shuttle. Would they ever be needed? The shuttle was designed with great equipment redundancy, so if something failed there would probably be a backup to take over. Also, it was feared astronauts could make a mistake and cause even more problems while trying to fix something. Fortunately, however, these engineers decided a tool box would be a small insurance policy against unforeseen failures and they assembled an in-flight maintenance kit that includes such common garage tools as screwdrivers, pliers, wrenches, wire cutters, hammer, voltmeter, tape, wire, and so forth. Also, NASA compiled a checklist that contains procedures for correcting some possible malfunctions (e.g., procedures on how to switch an operable computer screen in the back cockpit with a failed unit in the front, or remove a panel to get at a faulty switch). The kit has been used frequently to correct failures, and its use has become part of an astronaut's training program. Prior to flight, astronauts practice removing various panels and equipment in the cockpit of a real shuttle, so they are prepared for various maintenance tasks in orbit. There are also tools for spacewalks, for example, winches to hand-close the payload bay doors if their mechanical closing mechanism fails.

### When was a fly swatter used on the shuttle?

On mission STS-51D a large, expensive communication satellite did not electronically start as planned when it was released by the crew. Engineers on the ground suspected a switch on its side had not closed and wanted the astronauts to try and flip it. But the task seemed impossible. The satellite was spinning so it couldn't be approached by a spacewalker. NASA engineers, however, came up with a novel idea they dubbed the fly swatter. They directed the crew to use the stiff plastic covering of a checklist to fashion the swatter end. Slots were cut out of the center of the swatter and two space-walking astronauts taped this notched plastic to the end of the robot arm. (We always carry several rolls of common duct tape.) From inside the ship, another astronaut then maneuvered the arm so that the plastic swatter rubbed along the side of the spinning satellite. It was hoped the plastic slot would snag the faulty switch, pull it into position and activate the satellite. As it turned out, the problem wasn't with the switch and the crew of another shuttle flight, months later, had to use a special electronic box to correct the problem and send the satellite to geosync orbit. Nevertheless, the fly swatter was a clever idea and an example of Mission Control's ingenuity.

## What animals have flown into space?

The first animal to reach orbit was a Russian dog launched aboard the second Earth satellite, *Sputnik II*. The dog died when the satellite overheated. In America's space program, NASA opted to use primates as their early space biomedical subjects, and several species of monkeys and chimpanzees were launched into orbit. None were intentionally killed, but at least one died from overinvasive research (i.e., probes in the brain). Over the years there have also been many laboratory mice and rat astronauts. The shuttle has carried its share of animals including monkeys, rats, mice, and Marines (just kidding). The shuttle has also carried jellyfish, spiders, bees, frogs, brine shrimp, and gold fish. Many of these bug and animal astronauts were part of student experiments flown by NASA for schools. Space animal experiments are conducted to better understand how humans are affected by orbit flight. NASA researchers try to be as humane as possible, but some experiments require the subjects to be destroyed. This is particularly true with some of the shuttle rat-nauts—doctors want to compare the effects of weightlessness on various organs with a control group of rats left on the ground.

**Can astronauts carry their pets into space?**

No. The only animals that are carried are for NASA experiments.

**Could you walk a dog in space?**

Only if it was in a pressure suit.

**Could you fly a paper airplane inside a space shuttle?**

Yes. Astronauts have tossed paper airplanes around in the shuttle. But they don't fly as they do on Earth. Without gravity to counter the lift from the wings, the planes will always climb.

**Do astronauts carry toys into space?**

Yes, but not for personal recreation. Slinkys, paddle-balls, tops, jacks, and various other toys have been flown aboard the shuttle as a novel way of getting children interested in space. Videos of these weightless toy demonstrations are distributed by the NASA Headquarters Education Department.

# CHAPTER 4

---

# Life in Space

**How much room is there in a space shuttle?**

Not much. To better appreciate how tight the shuttle's quarters are, imagine being with five people (who haven't had a shower) for 2 weeks in a 100-square-foot room (the approximate living area of our mid-deck). If you want an even more vivid mental picture of the situation, imagine you are clamped to the toilet in a call of nature while somebody is 6 feet away preparing supper. I think I've made my point. The living area in the shuttle is small—about 100 square feet in the middeck and another 40 square feet in the aft flight deck. But, while it is tight, it's more livable than these numbers suggest. Remember, in weightlessness *all* of the volume is accessible. So, instead of being shoulder to shoulder with your crewmates, you could be floating in a ceiling corner (the middeck has a 7.5-foot ceiling). Weightlessness makes any "room" seem much more livable. Many people remark that the shuttle appears roomy in the TV shots that the astronauts send back. This is an effect of the wide-angle lens that is typically used on the cameras. They give false impressions of spaciousness.

**How many floors are there in the shuttle?**

There are two floors (called decks) to the shuttle cockpit. The upper deck is what most people would call the cockpit. It's where most of the switches, controls, and windows are. The downstairs deck is called the mid-deck, and it's the living room, bedroom, toilet, kitchen, gymnasium, and storage area for up to eight people (the maximum crew ever flown). There are two openings (one on each side) in the floor of the upper deck to allow astronauts to float between the two decks. On the port side, there is a ladder to get between decks when the shuttle is on the ground.

Underneath the floor of the mid-deck is a lower deck, but it's not a living space. It contains cooling duct plumbing, fans, pumps, a garbage volume, and so on. The lower deck volume is very crowded, and you can't stand or float vertically in it. It's like the crawl space beneath a house. Panels have to be removed from the mid-deck floor to get into the lower deck, and this would be done only in the event a malfunction required it.

## How many windows does the shuttle have?

Twelve. In the upper cockpit, six separate windows form the windshield. In the back of the upper cockpit are two ceiling windows and two aft-facing windows (that look into the payload bay). The only downstairs windows are small circular ones in the center of the side-entry and airlock hatches.

*View of shuttle windows as photographed by Russian cosmonauts aboard the Mir space station.*

## How thick are the shuttle windows?

Each of the shuttle's six forward and two overhead windows have three different window panes. The outermost pane is called a thermal barrier and is made of fused silica (similar to pure quartz) and is about a half-inch thick. On entry, it will have to withstand about 1,200° F of heat. The middle pane is 1.25 inches thick and serves as a pressure barrier. It's also made of fused silica. The innermost pane is another pressure barrier and is about a half-inch thick. It's made of tempered glass.

## How do you get inside a space shuttle?

There is only one ground entry to the cockpit of a shuttle—a port (left) side, mid-deck hatch. It's 40 inches in diameter, so you have to duck-walk or crawl on your hands and knees to get inside.

## Does the space shuttle have any emergency exits?

Yes. If a space shuttle crash landed and the side hatch was jammed (or there was a fire near it), the astronauts can pull a handle in the cockpit that will blow open the port-side overhead window (in the ceiling of the upper cockpit). The trapped astronauts can

then climb on the port-side mission specialist chair and lift themselves out of this window and onto the roof of the shuttle. Repelling cables, which are stowed in the cockpit, can be thrown overboard and used to repel down the side of the shuttle. (I found this to be one of the scarier training exercises we did at NASA. Trying to get off the sloping roof of a shuttle mock-up 20 feet off the ground while wearing 60 pounds of constricting pressure suit and helmet was definitely fear inducing.) For a discussion on bailing out of a space shuttle, see Chapter 7.

*An astronaut repels from the top of a shuttle in a training exercise.*

### Does the shuttle have a smoke alarm and fire extinguishers?

Yes. Few things are more fearful to an astronaut than an in-orbit fire (there's no way to escape it), and the shuttle has numerous fire protection features. The first line of defense is not to put dangerous materials in the cockpit, so everything aboard the shuttle is subjected to stringent flammability tests. Also, in areas where inaccessible electrical fires are a possibility (e.g., the avionics equipment bays), NASA has installed fire detectors and halon fire extinguishers that can be activated by a cockpit switch. There are also portable fire extinguishers that can be discharged through fire holes in instrument panels to attack fires that are hidden behind these panels.

### Do astronauts watch TV in space?

When I was an astronaut, there was no way to receive ground television signals aboard the shuttle. In some NASA photos you might see what appear to be TV sets in the cockpit, but these aren't used for ground TV reception. Instead, they are used to observe what the shuttle's own TV cameras are viewing. (There are TV cameras in the cockpit as well as in the payload bay and on the robot arm.) However, NASA is developing a capability for astronauts to watch TV signals sent from Mission Control. Such reception could be used as a morale booster by allowing two-way video conferences with family members on the ground or by transmitting a Super Bowl game or other video entertainment. Also, it would allow Mission Control to send video instructions on shuttle repairs or experiment operations.

### Do you get claustrophobic in a space shuttle?

No. In fact, the view from the space shuttle windows is so expansive and spectacular, you experience an opposite feeling—freedom.

*The earth seeen from overhead and aft-facing windows.*

### What do astronauts do for entertainment in space?

In space it's only necessary to look out the window to be entertained. The sights are so beautiful, it's impossible to get your fill of them in a mission as short as the typical shuttle mission (a week or two). Any time astronauts have a free moment, they are at the windows, staring in amazement. In fact, we carry cleaning material to wipe the nose prints from the glass.

There is one other source of entertainment—music. NASA provides everybody with a Walkman-style tape player or CD player, and astronauts are allowed to carry tapes and CDs of music of their choice. Most of the music I carried was classical (Beethoven, Pachelbel, Bach, etc.). Many astronauts listen to their tapes in their quiet moments: watching from the window and falling asleep.

Recently, NASA permitted the crew of a long-duration shuttle mission (STS-62) to carry several taped movies (and this will probably be an entertainment source on the future space station). The shuttle's camera equipment includes camcorders, and there are color monitors in the aft cockpit. The crew played the movies in a camcorder that was plugged into one of these monitors. There were no musicals aboard. The crew selections included *Terminator II* and *Predator II*. (I think it'll take a brave crew to watch the movie *Alien* in space.)

### Do shuttle astronauts ever get a day off while in space?

Yes. On missions longer than 10 days, an astronaut will be given a half day off work. While I never flew a mission that long, I suspect most astronauts would continue to do some work during that time. We all have a fierce need to be better than the last crew—to get more data, to take better photos, and so on.

### Does the space shuttle toilet flush with water?

No. Water won't flow in weightlessness, so it can't be used in the space toilet. Instead, our space toilet uses air. Basically, astronauts go to the bathroom in a vacuum cleaner.

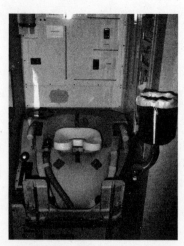

The shuttle toilet.

### How do astronauts urinate in space?

Refer to the photo on this page. The urinal is the hose that comes up from the bottom-front of the toilet (it is shown Velcroed to the side for launch). It's exactly like a vacuum cleaner hose. Urine is sucked through this hose and carried into a holding tank. When that tank is full (about every 3 or 4 days), the urine is dumped overboard. Urine dumps are beautiful to watch because the fluid instantly freezes into a glitter of shiny ice crystals. (I'm certain frozen urine is also the source of some UFO sightings alleged to have been accidentally recorded by shuttle TV cameras.) To accommodate the anatomic differences between men and women, the urinal has funnel adaptors. Men attach a standard, wide-mouthed funnel to the end of the hose before urinating. Women attach an adaptor that is anatomically shaped to permit a seal against the body.

### Are astronauts catheterized?

At no time does shuttle waste collection involve catheterization of astronauts of either gender.

### How is solid waste collected in space?

Solid waste collection is also done via air flow. There's a small hole (about 4 inches in diameter) in the center of the toilet seat. A flat cover slides in and out of the way to close and open the hole. Around the inside of this opening is a ring of small air-suction holes. Both the air-suction on–off function and the slider open–close function are controlled by the silver-knobbed lever on the right side of the toilet. The astronaut sits on the toilet and moves the lever forward. This opens the hole by sliding the cover out of the way and, at the same time, turns on the air suction. Cabin air is sucked through the holes and pulled downward into the toilet bowl taking the solid waste with it.

When the astronaut is done, he pulls the lever back, the slide cover closes the opening, the air suction is turned off, and the waste is trapped below. Solid waste is returned to Earth. Toilet paper can't be placed in the toilet. The silver can on the left side of the toilet has a vacuum cleaner-like bag for this purpose. Suction holds the used toilet paper inside. When done, the astronaut closes the bag and stows it in a container at the back of the toilet.

Perhaps the most amazing thing about astronaut training involves the solid waste collection feature of the shuttle toilet. The small size of the toilet opening (about 4 inches) makes *aim* critical to the success of the operation. Recognizing this, NASA engineers have built a toilet trainer at the Johnson Space Center with which astronauts can practice their aim. This trainer consists of an upward-pointing television camera set in the toilet bowl. Astronauts sit on this toilet and check their aim by viewing a television that sits in front of the trainer. When they have properly positioned their bodies, they memorize the position of their thighs and buttocks in relation to the toilet seat. To say this training takes a lot of the glamour out of being an astronaut is an obvious understatement.

### How do astronauts sit on the toilet?

In the previous answer, I said the astronaut sits on the toilet. Actually, it's impossible to sit on anything in weightlessness. Astronauts clamp themselves to the toilet. On each side of the toilet seat are things that look like handles. Prior to solid waste collection, an astronaut floats over the seat, pulls up on these handles (against a spring force), and twists them inward over their thighs. The spring force clamps the thighs and keeps the astronaut from floating away.

### How does an astronaut go to the bathroom when they are in a pressure suit?

There are three times when an astronaut will be in a pressure suit and unable to use the shuttle toilet: launch, reentry, and spacewalking. During these mission phases, crewmembers wear urine collection devices, or UCDs (no provisions are made for solid waste collection). There are two versions of these devices available. One type of UCD is only usable by male astronauts and consists of a waist-worn nylon bladder that has a one-way urine valve on its inside-front. A condom-like sleeve attaches the astronaut's body to the device. Urine passes through the sleeve, through the one-way valve and into the bladder. Women must use adult-type diapers that have been modified with more absorbent padding. Many men also elect to use the diaper because of its ease of donning and security. (While struggling into the shuttle seat, it's possible for the male UCD to come unplugged. If that happened, you could end up urinating in your pressure suit.) After arriving in orbit or after a spacewalk, astronauts remove their UCDs and stow them in the wet trash container.

**Does the shuttle toilet make the cockpit smell bad?**

No. There is an activated charcoal air filtration system that removes odors.

**Are there separate toilets for men and women?**

No. Everybody uses the same toilet.

**Is there a door on the toilet?**

There is a toilet door but it's not like a door on earth. The toilet area is so small you can't close yourself inside, so the open door serves as a wall. Curtains are then Velcroed into place to completely shield the occupant (including a ceiling curtain because part of the toilet area is open on top).

**Who cleans the toilet?**

During the mission, the toilet is cleaned by each person after use. After landing, it's emptied and thoroughly cleaned by ground personnel.

**What would astronauts do if the toilet broke while the shuttle was in orbit?**

NASA isn't going to end a space mission if the toilet malfunctions. Instead, they will direct the astronauts to use UCDs for urine collection (extras are stowed aboard for just such a contingency) and plastic *Apollo* bags for solid waste collection.

These latter devices get their name from the fact that they were the *primary* solid waste collection system for the *Apollo* astronauts. (*Apollo* capsules had no toilets.) An Apollo bag looks like a top hat with an adhesive strip where the hat rim would be (for attaching the bag to the astronaut's buttocks). *Apollo* astronauts were *real* men to have been shoulder to shoulder for 2 weeks and used these devices.

I can speak with some authority on the subject of broken space toilets. About half-way through my first mission, a heater failed during a waste water dump, causing a block of ice to form over the dump port and preventing us from using our urinal for the remainder of the mission. Unfortunately, this was before NASA began to carry the extra UCDs for urinal contingencies, so we had to use the Apollo bags as UCDs. This was a real challenge. You can't just urinate into a bag in weightlessness as you might do on Earth because, as the fluid impacts the bottom of the bag, it bounces straight back. We rapidly learned it was necessary to stuff the bag with a wash cloth or a sock and regulate the flow of our urination to equal the wicking (absorbent) rate of the cloth material. Going too fast would cause the fluid to splash. Going too slow would allow fluid surface tension to keep big globs of urine attached to our bodies. Judy Resnik, our lone female crewmember, was even more challenged by this requirement to use bags for urination. Such is part of the hidden glamour of being an astronaut.

## Do astronauts wear magnetic shoes?

The science fiction movies got this *waaaaaay* wrong. Remember the scene from *2001: A Space Odyssey* when a Pan Am (talk about getting it wrong!) spaceclipper stewardess *walks* down the aisle to deliver a meal to a passenger? She holds herself by using Velcro shoes. In other sci-fi movies you'll see astronauts using magnetic shoes or suction cups to hold themselves. Shuttle astronauts use none of these. Instead, they have a foot retention system that is the essence of simplicity: canvas loops taped to the floor with common duct tape. To hold themselves in place, shuttle astronauts merely slide their feet under the loops.

For space-walking astronauts, the boots of their suits have a small lip at the back of the heel that will lock the boot to foot-restraints—devices placed around the shuttle's payload bay where a spacewalker will have to do work. They are also sometimes put at the end of the robot arm, so an astronaut can be maneuvered by the arm operator.

## How do astronauts move their bodies in weightlessness?

In the cockpit, you rapidly learn to propel yourself with your fingers. It merely takes a light touch of a cockpit surface to send yourself flying. On spacewalks, astronauts use payload bay handrails to move around. Except for holding your body, legs are useless in space.

## What do astronauts eat in space?

The shuttle carries a variety of food that's preserved in a variety of ways. Some is ready to eat in the form of US Army *meals ready to eat* (MREs). Usually the MREs are main course items, like barbecue meat, spaghetti and meat balls, chicken ala king, and so on. Other shuttle food is dehydrated. It's similar to commercially available camping food and includes meats, vegetables, and deserts. Astronauts also have items of normal Earth food (bought right off a supermarket shelf), like pop-top cans of fruit and pudding, cookies and candy, peanut butter, and other canned sandwich spreads. For the first couple days, there might even be fresh fruit. Food preparation varies with the food type. MREs need only be heated in the galley oven. Dehydrated food takes the most preparation time. Individual food items are packaged in small plastic dishes that have a plastic covering to keep the food inside. On one corner of each dish is a small septum. The galley rehydration station has a water needle that inserts into this hole. Switch controls are then used to dispense the correct amount of either hot or cold water through this needle, under the plastic covering and into the dish. To facilitate rehydration, the astronaut stirs the food and water by running their fingers back and forth across the plastic cover. Then, if it's a hot food item, it can be further heated in the oven.

## How do astronauts eat in space?

**Astronaut eating M&Ms.**

Actual food consumption is not too different from Earth eating. Scissors are used to cut back the plastic cover on the dehydrated food dishes and to partially open the foil packages of the MRE's. (Astronauts never cut anything completely away because that creates a separate item of garbage that has to be controlled.) Then, forks and spoons are used to transport the food to the mouth. Most food items are sticky enough to stay in the containers and, with careful utensil movements, there are not too many spills. Occasionally, though, an astronaut will have to corral a floater with his silverware or hands and herd it back to his mouth. The trick is to eat slowly and not make any sudden motions.

NASA provides trays that serve as individual tables. These can be Velcroed to the astronauts' clothes to form a lap table. The dehydrated food dishes can be pressed into cut-outs in the tray, while a corner of an MRE package can be slipped under a tray clip. A magnetic strip holds silverware to the tray.

## Can you swallow food in weightlessness?

Yes. Swallowing is a muscular action that is completely independent of gravity. In fact, on Earth you could eat a meal standing on your head (which requires swallowing *against* gravity), so swallowing in weightlessness would be even easier.

## Do shuttle astronauts eat food out of squeeze tubes?

No. Squeeze tubes of food were used in the earliest days of space travel. Shuttle astronauts have MREs and dehydrated food.

## Can you eat food with a spoon in space?

Yes. Since the g-force that holds food in the dish of a spoon on Earth is absent in weightlessness, I could not imagine that a spoon would be functional in space, until I ate a can of fruit cocktail. What happened? As the spoon was drawn out of the can, the surface tension of the juice caused some of the fruit to stick to the spoon. The spoon transported the food to my mouth, just as it would have on Earth except that sometimes the fruit came to my mouth on the bottom of the spoon, or hovered on the tip or side of the spoon.

## Does space food taste good?

Shuttle food is not as tasty as regular Earth food, but it's okay for missions as short as the shuttle flies—a week or two. However, on a long-duration space mission (e.g., a space station mission or a Mars mission), a better menu is going to be necessary for crew morale. Frozen steaks, lobsters, ice cream, and other premium Earth food will be needed.

## Does food taste different in space?

Some astronauts find food that is seasoned to their taste for Earth consumption tastes somewhat bland in space. Is the sense of taste affected by weightlessness? Doctors can't be positive, but the zero-g fluid shift (see Chapter 5), which fills the head with fluid, might have some affect on taste, just as the head congestion of a common cold will alter taste. NASA dietitians offer a menu with a variety of extra-spicy foods for any astronaut who feels their sense of taste might be muted by weightlessness. The galley also has additional spicy sauces that can be squeezed onto a meal.

**Can astronauts eat sandwiches in space?**

Yes, but the preferred bread is flour tortillas. In weightlessness it's much easier to spread peanut butter and other sandwich spreads on a tortilla than on a piece of ordinary wheat bread. The latter crumbles apart too easily and generates too many floating crumbs. Ingredients are rolled up in the tortillas to form the sandwich.

**Can astronauts pick their own menus?**

Yes. Preflight, everybody has a chance to sample the food and choose their own menu from a list of breakfast, lunch, and dinner items. Individual food packages are labeled with colored dots so it's easy to distinguish who has what packages. There is also a general pantry of food items (candy, cookies, nuts, etc.) that anybody can eat from at any time.

**How many calories does an astronaut eat daily?**

This is a function of the astronaut's gender, age, and size and varies from about 1,900 calories for a petite woman to 3,200 calories for a large male astronaut (a good reason why small women would be more efficient long-duration space travelers).

**Can astronauts salt and pepper their food in space?**

Yes, but not with shakers, as is done on Earth. If you tried to shake salt on anything in weightlessness, it would float everywhere. Instead, the shuttle galley contains small, squeezable bottles (like eye drop bottles) containing salt water and pepper water. If an astronaut wants to season a dish, she merely squeezes drops of the salt and/or pepper into her food.

**Do astronauts play with their food?**

Absolutely. I once built a solar system using a blob of orange juice as the sun and M&Ms as the planets. Also, I've played M&M

baseball with other crewmembers, hitting the candy with rod-like tools and catching them in the mouth.

### Do astronauts eat astronaut ice cream in space?

Though it is sold as a novelty in many museums (and advertised as astronaut ice cream), astronauts do not eat freeze-dried ice cream.

### Is gum carried aboard the shuttle?

Yes. Gum is a menu item. Chewing gum in orbit is no different than on Earth.

### Does the shuttle have a refrigerator or freezer for food storage?

No. On some missions there might be a freezer, but it won't be used for food. Instead it will be used to preserve various medical specimens. The one exception was STS-74 (a mission that docked with the Russian *Mir* space station), which had a freezer to return some *Mir* specimens that wasn't being used during launch. So NASA filled the freezer with real ice cream, which the astronauts shared with their Russian counterparts.

### Does the shuttle have a microwave oven?

No. It is feared such a device would interfere with the shuttle's electronic equipment. The shuttle oven uses electricity to heat the walls and ceiling, and a fan circulates air so items held in interior trays will get warm.

### Do astronauts eat pizzas in space?

No. Pizzas can't be dehydrated and they would take up too much room if they were taken in the form of normal Earth pizza.

**What was my favorite space food?**

Chocolate pudding. We carry cans of store-bought puddings (chocolate, vanilla, tapioca, etc.).

**Do astronauts put on weight in weightlessness?**

In space, astronauts can eat as much as they want without gaining an ounce of weight. Sounds too good to be true, doesn't it? And it is. While it's true you can't gain weight in weightlessness, you can gain mass. When you return to earth, that extra mass will show up on the bathroom scale as extra pounds. That's the difference between mass and weight. Mass is independent of gravity, so it's always with you, even in weightlessness. You can think of mass as a measure of the amount of stuff something is made of. Weight, on the other hand, only exists when that mass is under the influence of a force, like gravity.

When they return to Earth's gravity and stand on a scale, some astronauts will find they have put on weight, but most will not. The average astronaut loses a couple pounds during a week-long shuttle flight. For some astronauts, space sickness is responsible for this weight loss. But anybody can have their appetite suppressed by the sheer excitement of flying in space. An apparent altered sense of taste due to the physiologic affect of fluid shift (see Chapter 5) and the time associated with preparing a meal are also factors that reduce caloric intake. (Who wants to spend a lot of time preparing a meal when you could be looking out the window and snacking on peanuts?)

**Who does the cooking and dishwashing on the shuttle?**

Everybody. There is no designated cook or dishwasher. Meals are prepared by whoever is most available to do so. Afterwards, the food containers are thrown away so there is no dishwashing. Only the silverware is cleaned, and that's done by each astronaut with an alcohol wipe.

**How do astronauts drink in space?**

Somebody trying to drink from a cup or glass in weightlessness is going to be very frustrated. Nothing will pour in zero-g, so a turned up glass won't bring any fluid to your mouth. Astronauts must drink from straws. Our drink containers are aluminum pouches with powdered drinks inside and a fill port at one end.

Using the water needle of the galley food rehydrator, astronauts can fill these pouches with 8 to 12 ounces of hot or cold water (depending upon the beverage). After the pouch is filled, a plastic drinking straw is inserted into the fill port. A clip on the straw prevents fluid from escaping when the container isn't being used. Like the food containers, the drinks are color coded by crew position.

Astronauts can select from a wide variety of drinks for their menus—plain water, tea (sweetened or unsweetened, hot or cold), cocoa, coffee (cream/sugar/black), and an assortment of commercially available fruit-flavored beverages.

**Where do astronauts get their water?**

Our electrical power system (EPS) combines liquid hydrogen and liquid oxygen in devices called fuel cells to produce electricity. The waste product of this process is water that astronauts use for drink and food preparation.

**Do astronauts drink recycled urine?**

Urine is not recycled on the space shuttle, but it will be recycled to drinkable water on the space station.

**Do astronauts drink Tang in space?**

Yes. Early in the space program, Tang was selected as an astronaut beverage because it was nutritionally well balanced, came in multiple flavors, and was readily available.

### Do astronauts drink soft drinks in space?

No. Soft drinks (Coca-Cola and Pepsi) have been flown on the shuttle but only for evaluation purposes. They are not available as a regular menu item. Storage volume is an impediment to flying such ready-made drinks on the shuttle, but NASA is continuing to investigate the possibility of flying soft drinks aboard the space station.

### Do astronauts drink beer, wine, or other alcoholic beverages in space?

No. A French astronaut flying aboard the shuttle carried small bottles of wine, but these were not consumed in orbit. Instead, they were returned to Earth and given away as gifts.

### Do astronauts smoke cigarettes in space?

No. While a handful of astronauts have been smokers, nobody is allowed to smoke on a shuttle. There are rumors, however, that some Russian cosmonauts smoke on their *Mir* space station.

*Author sleeping.*

### How do astronauts sleep in space?

There's no such thing as lying down in weightlessness, so there are no beds in a space shuttle. Astronauts can sleep anywhere, in any orientation, because everywhere and every orientation is identical—weightless. Theoretically, a sleeping shuttle crew could freely drift in the cockpit like motes of dust. Such free drift sleep isn't practical, however, because people would be constantly waking up as they floated into each other or into contact with the cockpit boundaries. Also, a crewmember on the way to the toilet wouldn't like having to weave through bodies to get there. So, most astronauts use sleep restraints. These are merely cloth bags that can be clipped to the walls or ceiling. Astronauts zipper themselves into these bags to keep from floating around and to keep warm while they sleep. Other astronauts forgo the sleep restraints and merely strap themselves into the commander's or pilot's seats. Usually, though, sleep in the upper cockpit is difficult. Approximately every 45 minutes the sun will rise and its light and warmth will disturb a sleeper.

The most bizarre thing about weightless sleep is that astronauts' arms float in front of their bodies when they're unconscious. The first time I awoke in the middle of a sleep period and saw my fellow crewmembers asleep on the wall with their arms gently waving in front of their bodies, I thought I was in a science fiction movie.

On missions where there are round-the-clock operations, usually *Spacelab* missions, the sleep restraint is impractical. The working shift of astronauts would find it impossible to be so quiet as to not wake the sleeping shift. On these missions, NASA installs four, coffin-like berths on top of each other on the starboard side of the mid-deck. Each berth has a sliding entry door that can be closed to ensure isolated sleep.

**Astronauts in sleeping berths.**

### How much time do astronauts get to sleep?

Each sleep period is approximately 8 hours, though few astronauts actually get that much sleep. Most will stay up past bedtime to look out the windows.

### Do astronauts dream in space?

In all of my space sleep periods I never dreamed, but others have. One astronaut told me that on one of his missions he had a nightmare that caused him to wake up shouting. As you might imagine, his shouts had everybody else on the crew in a momentary panic.

### Is is possible to snore in space?

While I never heard anybody snoring on any of my missions, some astronauts have heard fellow crewmembers snoring. There is certainly nothing unique about weightless sleep that would eliminate or aggravate snoring. If you do it on Earth, you'll probably do it in space. One thing is certain about snoring in weightlessness. Rolling over isn't going to stop the noise. There's no such thing as rolling over in weightlessness. Every position is identical.

### Do astronauts exercise in space?

Yes. Generally, the flight plan will reserve an hour each day for every astronaut to exercise. The mechanism for this exercise is typically a stationary bicycle device, but NASA has also flown rowing-type exercisers and treadmills (bungee cords hold you to the treadmill to simulate gravity and allow you to run). The importance of exercise is a function of the duration of the mission. The shorter the mission, the less profound the physiological effects of weightlessness. For missions less than 13 days in duration (most shuttle flights), flight surgeons require that exercise periods be scheduled, but whether the crew actually does exercise is optional. For missions longer than 13 days, the doctors require periodic

**Astronaut exercising.**

exercise to keep the body's cardiovascular system conditioned and to minimize the potential for fainting on reentry and landing.

When exercise is optional, there are two reasons many astronauts (myself included) skip it. First, an hour of exercise is an hour of window time that's lost. Earth watching is a powerful distraction and makes it easy to play hooky from the treadmill. Second, getting hot and sweaty when there's no shower to step into doesn't make a lot of sense.

### How do astronauts brush their teeth in space?

Astronauts brush with regular toothbrushes and their choice of toothpaste. The only difference in weightless tooth brushing is that there's no sink in which to spit the foam, so it must be spit into a tissue.

### Can you wear glasses or contact lenses in space?

Yes. Normal Earth glasses work fine in space. The squeeze of the eye-glass frame against the temples is enough to keep them in place. Also, many astronauts wear contact lenses in space. Since these are fitted to the individual's cornea, their vision-correcting function is independent of gravity's pull.

### Can you put eye drops in your eyes in space?

Yes, but you must touch the drop of liquid to the surface of the eye so it will disperse over the eye. Of course you could also float a drop of liquid in front of your face and move your open eye into it, but I've never seen anybody try this.

### Does the shuttle have a sink for washing hands and other hygiene activity?

No, water will not flow in weightlessness, so there would be no way to hold your hands under a faucet as you do on Earth. We use alcohol wipes or a wet wash cloth and soap to clean our hands.

## Can you take a bath or shower in space?

On the 1974 *Skylab* missions American astronauts had a space shower. They wet their bodies from a hose while a curtain surrounded them to prevent droplets from floating away. Excess water was vacuumed away. There's no room on a shuttle for such a device, so shuttle astronauts don't take showers. The only way a space shuttle astronaut can wash is with a wet wash cloth and soap. To shampoo their hair, they use rinseless shampoos (as are used for bedridden hospital patients). In my opinion, the lack of a shower is the worst aspect of flying in space.

## Do male astronauts shave in space?

Yes. They can select either a battery powered electric shaver or wet-razors. Most use the electric. Those who do wet shave have no sink in which to wash off the razor, so they must wipe the blade with a tissue.

## Do female astronauts shave in space?

Yes, some female astronauts shave their legs and underarms in space. Of those who do, some use a wet razor, with the same difficulties the males experience. Others use an electric razor. For both sexes, personal grooming in space—bathing and shaving—make you feel fresher, just as on Earth.

## Do female astronauts wear makeup in space?

Some do. Some don't. NASA does have a shuttle makeup kit consisting of a base, blush, mascara, lipstick, and eye shadow.

## Can astronauts paint their fingernails in space?

No. While there is no reason fingernail painting wouldn't work in weightlessness, nail polish is flammable and therefore too dangerous to carry.

**Are feminine hygiene products put aboard the shuttle for female astronauts?**

Yes. Every individual has his or her own locker drawer. Lockers for female astronauts contain their preferences in hygiene products.

**Have female astronauts had their menstrual cycle while in space?**

Yes. These women have reported no difference in the timing, duration, or flow of their cycles.

**Where are things stowed aboard the shuttle?**

Hundreds of pounds of food, clothes, checklists, tools, cameras, medical equipment, experiments, and so on, must be packed so as to be easily accessible during flight. The primary stowage locations are mid-deck lockers, covering the entire front wall of this part of the shuttle cockpit. Each locker has a door that can be folded down to expose trays of equipment. A checklist cross-references equipment items with their stowage locker. A net covers each tray so when it's pulled out, items won't float everywhere. You reach under the net to get whatever you want.

*Author looking in shuttle locker. Note that my feet are being held under canvas foot loops.*

**Can astronauts set down things in space?**

No. Items must be restrained or they will float away. The cockpit is covered with small pads of Velcro pile material and most items have a Velcro hook glued to them so they can be stuck on a shuttle surface. One of the difficulties about this arrangement is that you tend to forget where you've put something and end up wasting time searching for it. Unlike on Earth, where things are always put *down*—on the floor, table, countertop, desk—in space, you can unthinkingly put items *anywhere*. Besides Velcro, astronauts also use string-like tethers to keep checklists from floating away.

### Where do astronauts put their garbage?

There are separate stowage areas for the two types of garbage—wet and dry. Wet garbage, which includes food waste, used UCDs, and emesis (barf) bags, is stowed under the mid-deck. It's accessed through a small trapdoor in the floor. A flexible rubber covering allows new garbage items to be shoved inside while preventing previously stored items from floating out. While this garbage pit contains items that have an unpleasant—if not nauseous—smell, the cockpit air conditioning system keeps those odors from entering the cockpit living area. Dry trash (mostly paper items) is stowed in canvas bags in each cockpit and in a small, wall-mounted volume in the mid-deck.

### Does the shuttle have a trash compactor?

A compactor is available if the crew wants it aboard their mission. However, most crews feel they can manually control the garbage volume as well as the compactor could. The compactor takes the place of valuable storage locker space. Many crews would rather have the extra locker, so they don't opt for the compactor.

### Does the shuttle have a vacuum cleaner?

Yes. A hand-held, portable vacuum cleaner (modified with a muffler) is carried on shuttle missions. Its primary use is to vacuum lint and other debris from the air filters.

### Who does the shuttle housekeeping?

Some housekeeping tasks are scheduled in the flight plan and assigned to a specific crewmember (e.g., cleaning the air filters and replacing lithium hydroxide canisters used to absorb carbon dioxide). The rest of the housekeeping is a shared responsibility and most astronauts make a significant effort to ensure the shuttle is kept as clean as possible, even to the extent of scrubbing the walls. (After several days of occupation, there will be dried spots of spilled drinks and possibly even small specks of dried vomit on the cockpit surfaces.)

### What does an astronaut wear inside the shuttle?

After reaching orbit, astronauts strip off their pressure suits and UCDs and don cotton golf-type shirts (long and short sleeve) and gym-style shorts. On many missions the shirts are Lands' End products monogrammed with the orbiter name and mission number, but the crew has wide latitude in ordering their shirts from any retail company. The shorts have strips of Velcro across the thighs so oversized, portable pockets or a lap food tray can be attached. Astronauts load the pockets with pens, pencils, note pads, eating utensils, tape recorders, and so forth.

### Do astronauts have a change of clothes?

Yes. There is one pair of underwear for each mission day and a couple extra sets of shirts and shorts.

### Is there a dirty clothes hamper on the shuttle?

Yes, in the form of a canvas bag. It's usually tied in the airlock along with net bags containing the launch/entry pressure suits.

### Do women wear bras in space?

Some do. Some don't. In weightlessness they are obviously not needed for support.

### Do astronauts wear shoes in the shuttle?

No. There's no such thing as walking in weightlessness so shoes are not functional. Instead, we cover our feet with slipper-socks.

## How much privacy is there aboard a shuttle?

On missions with a male-female crew, privacy is more a function of behavior than design. My first mission, STS-41D, had a woman aboard (Judy Resnik), and the men would stay in the upper cockpit whenever she wanted to change clothes. If such an arrangement isn't practical, it's always possible to slip behind the curtained toilet area for privacy.

## Can Mission Control watch the crew with the cockpit cameras anytime it wants?

Besides the possibility of a fellow crewmember invading someone's privacy, there's also the potential for a camera to put someone's naked backside on world TV or to catch them with their faces in a barf bag—not exactly Right Stuff images. For this reason, astronauts are always very conscious of which TVs are active. Mission Control is also sensitive to astronaut privacy and will not activate any in-cockpit TV camera without first asking permission. If there's any doubt about TV coverage, astronauts have a cockpit switch to block transmission of all TV signals.

## Can an astronaut have a private conversation with a NASA doctor?

Yes. A private medical conference between a NASA flight surgeon and the crew is scheduled in the flight plan at the end of each work day. Also, a crewmember can request a medical consultation at any time, and that conversation will be blocked from the press. Only if a medical problem is determined to have a mission impact will NASA make a public statement on it.

## Is it hot or cold inside the space shuttle?

There is a knob on one of the commander's panels that can be adjusted to regulate the temperature—similar to setting a thermostat in a house. And, just as occurs in any home, there are complaints among individual crewmembers that the temperature is too hot or too cold. Generally, these complaints are along gender lines. When the female astronauts are comfortable, the male astronauts think it's too hot. When the male astronauts are comfortable, many of the women complain it's too cold.

## What do astronauts breathe?

The shuttle atmosphere is identical to the earth's sea level atmosphere: approximately 80% nitrogen and 20% oxygen at a pressure of 14.7 psi. The oxygen is carried in liquid form (in the same tanks that supply the electrical system fuel cells) and heated to a gas to replace metabolic consumption. The nitrogen is carried in gaseous form in high-pressure tanks. The atmosphere is continuously circulated through replaceable lithium hydroxide canisters to remove carbon dioxide.

## Do astronauts use pencils and pens in space?

Yes. The pencils are of the mechanical type. The ink-cartridges of the ballpoint pens are gas pressurized so the ink will flow in a weightless environment.

## What's the hardest thing to get used to while living in weightlessness?

With the exception of the physiological adaptation problems (back ache and vomiting), I think most astronauts would agree it's the additional time overhead associated with weightless work that's most difficult to cope with. For example, a simple 2-minute Earth task like changing the batteries in a recorder can become a 10-minute job in weightlessness. Used batteries, new batteries, the recorder, and the battery cover might end up floating in different directions. For many tasks you wish you had four hands. Some tasks, like meal preparation, eating, and solid waste toilet operations, can easily exceed by a factor of 10 the time required in comparable Earth activity.

**Is it fun to fly in space?**

Yes! Yes! Yes! Even the discomfort of the physiological changes, the nuisance of using a vacuum cleaner for a toilet, the not-so-appetizing dehydrated food, and the lack of a shower can't diminish the boyish and girlish glee astronauts experience in space. For many astronauts, spaceflight is a lifetime dream come true, and it's impossible not to feel an overwhelming rush of joy when you look from the cockpit windows. The beauty of the Earth from orbit heights is something difficult to capture in words, but the soul has no difficulty in celebrating the experience. It's your wedding night, your children's births, your first Little League home run, and every childhood Christmas, rolled into one.

# CHAPTER 5

# Space
# Physiology

## Does spaceflight change your eyesight?

There is a persistent story among space enthusiasts that astronauts develop super-vision in weightlessness and are able to see with an acuity far exceeding Earth vision. This is not true. Astronauts have done many vision tests in space and their visual acuity remains the same in orbit as on Earth. The super-vision stories probably developed when early astronauts first reported seeing human-made features like the Great Wall of China and highways and runways. Since most people thought these astronauts were orbiting at an *extreme* distance from Earth, they probably assumed astronauts' vision must have gotten significantly better in space. The fact is, astronauts see these objects because their vantage point (i.e., an orbit of about 200 miles high), is not an extreme distance at all. It's about the horizontal distance you can see from a high-flying airliner.

## What's it like to bleed in space?

I once saw an astronaut bleed, so I can answer this question as an eyewitness. The astronaut—suffering from space sickness—was receiving an injection of an anti-nausea drug. While the needle was still in his skin, he moved, causing a tiny skin tear. Blood came from the wound, but because it was weightless, it didn't run down his leg as it would have on Earth. Instead, it bubbled into a ball. When he moved, the ball separated from his skin and floated in the air like a red marble.

## What's it like to sweat in space?

The only time an astronaut will develop a space sweat is while rigorously exercising or possibly on a spacewalk. When such sweating occurs, it's just like bleeding. The sweat comes out of your pores, but it doesn't run. It just stays on your skin as tiny balls of water.

## Can you cry in space?

Yes. I cried on one of my space missions, overcome with emotion at the sight of my childhood home, Albuquerque, New

Mexico. It was while growing up in Albuquerque that I had first dreamed of being an astronaut.

What happened to my space tears? Just as with sweat or blood, tears don't run in space. They just cling to your eyes and eyelashes.

### Can astronauts get the hiccups in space?

Yes. Though I didn't get them on any of my missions, there is nothing about weightlessness that will change a person's susceptibility to hiccups.

### Do astronauts get motion sick in space?

No. While about 40% of astronauts vomit in the early days of a mission, this sickness is not the same as Earth-based motion sickness. There have been many cases of astronauts who are prone to terrestrial motion illness (seasickness and air-sickness) but who have never felt ill in space (I was in this category). Then, there have been many astronauts who have never been motion sick on Earth but have vomited for 2 or 3 days in space. So you cannot describe the sickness as motion sickness. The newspapers frequently make this mistake. Whatever causes the illness remains a mystery to NASA researchers. Some people are over the sickness in a day. Most people are well within 3 days. On rare occasions, astronauts have been known to be sick for a week.

### Is there a treatment for space sickness?

While they don't know the cause, the symptoms of space sickness can be treated. NASA flight surgeons have found the drug *Phenergan* to be very effective at stopping the malaise, vomiting, and other symptoms of the sickness, so this drug is carried aboard the shuttle. Space-sick crewmembers can get an injection of Phenergan from the crew member who acts as doctor. The downside of this treatment, however, is that the drug tends to make people drowsy. The ear-patch anti-nausea medicine that people use for seasickness is ineffective in stopping space sickness.

## What do astronauts use as a vomit receptacle?

The space toilet is not designed for emesis (vomit) collection, so astronauts must use plastic barf bags. As one might imagine, this a mess in weightlessness. When the emesis hits the bottom of a barf bag, it comes directly back into the face of the astronaut. Spills are frequent, and on some occasions people have been known to vomit before they have been able to get a bag to their faces. This is why rookie astronauts—who are uncertain of their susceptibility to space sickness—will usually spend their first day floating around with a barf bag half exposed from a pocket so it can be reached in a quick draw if the illness strikes them. Used barf bags are stowed in the wet-trash container that is underneath the floor of the mid-deck.

## Could an astronaut get sick on a spacewalk?

Yes, and this could be life-threatening. A spacesuited astronaut would have no way to get the fluid away from his face. He could aspirate it and choke to death. Also, the emesis could clog up the oxygen circulation system and suffocate him. This is one reason why NASA never schedules a *planned* space walk any earlier than the third day in orbit. This allows astronauts time to get over the sickness.

## If astronauts get space sick, are they grounded from future missions?

No. Early in the space program, astronauts tried to hide instances of inflight sickness because they didn't understand it and thought flight surgeons would ground them. As a result, doctors and other researchers had a difficult time collecting the data needed to understand the phenomenon and develop treatments for it. In the shuttle program NASA managers made it clear space sickness would have no affect on flight status (or even on spacewalk status), and doctors now get cooperation instead of deception.

**Do astronauts grow taller in space?**

Yes. I'm 5 feet 9 1/2 inches tall on Earth, but in space I grew to 5 feet 11 inches. In weightlessness, the vertebrae of the spine spread apart causing this 1- to-2-inch height growth.

Unfortunately there's a bad side effect to this phenomena. The lower back muscles do not stretch quickly to make room for this spine lengthening, and the result is a very painful back ache. Almost all astronauts complain of a bad back ache for the first week of space flight. Personally, I found this to be the most bothersome aspect of the body's adaptation to weightlessness. (People who are afflicted with nausea probably would say *that's* the worst adaptation syndrome, but I had no nausea.)

Besides treating this pain with pills (aspirins, etc.), astronauts can temporarily reduce it by rolling into a ball and pulling their knees toward their chests. It's not uncommon to see astronauts take a break from their experiments and perform this curling maneuver. Also, the astronaut sleep restraints have wide Velcro straps across the front of the bags. These allow an astronaut to "Velcro" themselves into a curl, thus reducing their back pain and helping them to fall asleep.

**Are spacesuits made bigger to make room for growing astronauts?**

Yes. NASA engineers anticipate the inflight spinal growth when they assemble the spacesuits for the extra-vehicular activity (EVA)-designated crewmembers. The suits used in space are taller than the suits used in Earth EVA training.

**If astronauts float upside down, does the blood rush into their heads?**

No. In weightlessness *every* attitude is identical. There is no such thing as upside down. With their eyes closed, astronauts have no idea where the ceiling and floor are located.

But, while body orientation has nothing to do with blood rushing into an astronaut's head, all astronauts do feel as if they are standing on their heads. The reason for this sensation is a physiological effect labeled by NASA doctors as fluid shift. Here on Earth, gravity keeps a lot of fluid trapped in our legs. In a couple hours of weightlessness, however, this fluid is equally distributed throughout the body. Legs get skinnier, while the upper body swells. Everybody's chests—male and female—get bigger in weightlessness. When I looked at myself in the mirror, I looked as if I had been pumping iron. All the muscles in my upper body were bulging. But the fluid shift doesn't stop with this Arnold Schwarzenegger look. It also fills the astronaut's heads with fluid and gives them a sensation similar to being upside down on Earth. (But, remember, the fluid shift occurs because of weightlessness, not body orientation.) Look carefully at pictures of astronauts in space and their faces will be puffy from this fluid shift. The head fullness is mildly uncomfortable, but you get used to it.

### Does your hair float in space?

Yes. Weightlessness makes every day a bad hair day for astronauts with long hair. It floats around their heads as if it's been wildly teased. For this reason, many female astronauts use clips or ties to bind their hair when they're in space.

### Do astronauts' bones get weaker in space?

Yes. Calcium and other minerals are lost from bone mass whenever it's not being stressed by gravity (the same thing happens to bedridden people), so

*Weightlessness makes every day a bad hair day for astronauts with long hair.*

weightless astronauts' bones do get weaker. This decalcification isn't significant on week-long shuttle missions, but it is a concern for long-duration space missions. NASA doctors are investigating ways of inhibiting the loss while in space.

## Does an astronaut's heart get weaker in space?

There is some evidence of minor decreases in heart pumping power after long-duration missions (month-long flights), but no significant changes have been seen after short-duration shuttle missions.

## Are astronauts in danger from space radiation?

Yes, but the danger isn't great. Shuttle astronauts remain in low Earth orbit (100 to 400 miles) so they are somewhat protected from significant radiation by the earth's magnetic field. On average, each day in space exposes an astronaut to the same radiation someone on the ground will receive each year from naturally occurring background radiation. While that might sound significant, a space dose is still only *thousandths* of what scientists feel a human can absorb without danger of long-term ill effects.

It is possible for an astronaut to be subjected to much greater doses of radiation during periods of high solar activity (solar flares). If such a threat ever developed, however, Mission Control might elect to bring the astronauts back early (or delay the launch if solar flares are predicted). Also, because they will be flying for long durations without the protection of the earth's magnetic field, Mars-bound astronauts will face a much greater danger from space radiation, and appropriate shielding will have to be included in the design of their craft.

## Does spaceflight change blood pressure?

You bet. Put yourself on the top of a rocket and see what happens to *your* blood pressure! In a serious vein, though, weightlessness has no affect on blood pressure.

## Do astronauts get younger when they fly in space?

No. But based on Einsteinium relativity, an astronaut will age *less fast* while in orbit than someone remaining on the Earth for the same time span. During a week-long space mission, an astronaut will age a fraction of a *second* less than she would have, had she remained on Earth. Such a change is due to the speeds at which an astronaut is flying. While it's nowhere near the speed of light (approximately 186,000 miles per second), orbit speed (nearly 5 miles per second) is sufficiently fast to cause this aging slow-down per Einsteinium relativity. A physicist friend calculated I am now 4 *ten thousandths* of a second younger than if I had remained on Earth because of my 2 weeks of space travel. (Regardless of what Professor Einstein's theory says, I never felt I was aging slower while riding a 4-million-pound bomb. In fact, I felt spaceflight aged me well beyond my Earth years).

### Do your ears pop in space?

The only time your ears pop in space is when the air pressure in the cabin changes. On a normal mission the pressure is kept at 14.7 psi from liftoff to landing, and so they don't. On a mission with a spacewalk, though, the air pressure does change, and so your ears do pop. Approximately 24 hours before the spacewalk, the cockpit cabin pressure is decreased to 10.2 psi to reduce the chance that the space-walking astronauts will get the bends. (Bends are a result of nitrogen gas coming out of solution in our bodies as the surrounding pressure is reduced. Because the space-walkers eventually will be exposed to a significantly reduced suit pressure of 4.3 psi, allowing them to breathe air at 10.2 psi for a full day before reduces the nitrogen in their bodies and thus reduces their risk of having the bends.) After the spacewalk, cabin pressure is increased back to 14.7 psi. As these changes in air pressure occur, astronauts' ears pop.

# CHAPTER 6

# Space Shuttle Reentry and Landing

## How does a space shuttle get out of orbit?

Deorbiting a shuttle requires that the vehicle be oriented tail forward and the *Orbital Maneuvering System* (OMS) engines fired for approximately 2.5 minutes. This will slow the craft by about 200 mph. This isn't much, when you consider it's still traveling about 17,000 mph, but the slow-down is enough to change the low point of the orbit so it will intersect the atmosphere. When that happens, atmospheric friction will do the real braking. After the deorbit burn, the shuttle is turned forward and its nose is raised to about 40 degrees above horizontal in preparation for hitting the atmosphere belly first.

## Could a space shuttle skip on the atmosphere?

Yes. If it flew too shallow of a reentry angle, it could skip on the air (like a flat stone can be made to skip on water ) and fly far beyond its intended landing site. On the other hand, too steep a reentry angle would be like hurling a flat stone at too great an angle into water. The stone will dig in and sink. Similarly, the shuttle would dig into the thick part of the atmosphere and be torn apart by high g-loads. This is why the shuttle guidance system has to keep it within a few *tenths of a degree* of the correct trajectory.

## Can a shuttle land anywhere?

Normally, a shuttle will be targeted to land at either Kennedy Space Center, Florida, or Edwards Air Force Base, California, but Mission Control has certified the deorbit target information for a handful of foreign airports and for the following other US sites. Who knows? Maybe someday a shuttle will drop in for a visit to your hometown.

White Sands, NM
Moses Lake, WA
Honolulu, HI
Ellsworth Air Force Base, SD

Nebraska Air National Guard Base, Lincoln, NE
Mountain Home Air Force Base, ID
Cherry Point Marine Corps Air Station, NC
Dover Air Force Base, DE
Myrtle Beach Air Force Base, SC
Oceana Naval Air Station, VA
Otis Air National Guard Base, MA
Pease Air National Guard Base, NH

The only place a shuttle has ever landed besides the Kennedy Space Center and Edwards Air Force Base is White Sands, New Mexico (STS-3). In this instance, the weather in Florida was unsatisfactory and heavy rains in California had flooded the normally dry lakebed of Edwards Air Force Base. Since the shuttle was not yet certified to land on the cement runway of Edwards, the only alternative was to land on the White Sands dry lakebed.

Mission Control can also uplink the deorbit targets for various foreign airfields around the world, but it would have to be an extreme emergency—a cabin pressure leak, a fire, a life-threatening illness, and so on—to require a landing outside of the United States. Many foreign airfields don't have the precision landing aids to assist astronauts in their approach. Also, the expense and difficulty of returning a shuttle to the United States from a foreign country would be significant.

## Why do some shuttles land in Florida and others in California?

On almost all missions, the primary landing site is the Kennedy Space Center. Unfortunately, the weather in Florida doesn't always cooperate, and the shuttle is forced to land in California. Rain is the biggest threat. It will destroy the shuttle's heat tiles, so if the landing forecast calls for storms in the vicinity of the Kennedy Space Center, NASA will direct the astronauts either to stay in orbit another day or to land at Edwards. Because it's located in the Mojave Desert, the weather at Edwards is almost always satisfactory for a shuttle landing.

There are missions in which Edwards Air Force Base is the

*primary* landing site. These usually involve some type of test of the shuttle's brakes, drag chute, or nose wheel steering mechanism. If a tire blows, in a brake test, for example, and the shuttle swerves off the runway, it's no big deal at Edwards, since the lakebed terrain is relatively flat all around. A swerve off the Kennedy Space Center runway would destroy the shuttle, since this runway is bordered by drainage ditches.

### Does the shuttle have any engines running during reentry and landing?

No. In every sense of the word, a landing shuttle is a glider from the moment its deorbit burn is complete. That occurs almost exactly on the opposite side of the earth from the planned landing site. Literally, a shuttle glides half way around the earth, across the Indian Ocean, Australia, and the Pacific Ocean—12,000 miles—on its way to a landing. At the end of that journey, its commander will get just one chance to land.

### How can the astronauts be sure a gliding shuttle will make it to a runway?

As soon as the deorbit burn is over, one fact is certain—the shuttle is coming to earth. The question is, will there be a run-

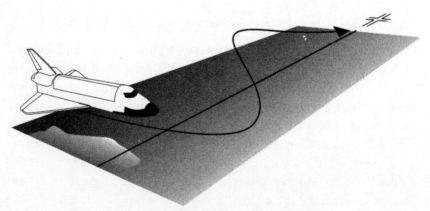

**The shuttle traces a giant "S" across the ground during reentry.**

way underneath it when it finally runs out of altitude? It seems an impossible problem: over a 12,000-mile, 1-hour glide, it has to arrive over a 3-mile long runway at the correct speed and altitude to land. And it has to do this without knowledge of many atmospheric variables that could severely affect the glide distance (e.g., upper altitude density and winds). How does it do it? Actually it's very simple.

The deorbit burn always leaves the shuttle on course but high on glide path. This is intentional, because it can always lose excess altitude, whereas, if it ever gets too low, it can never climb back up. (Remember, it's a glider.) So, the shuttle will remain high on glide path in a straight-ahead fall until it reaches about 65 miles altitude. At this point there is finally enough air to produce a small amount of lift on the wings. Now the shuttle can do something about being too high. It can't just dive straight down to the correct altitude because the increased speed would burn it up. Instead, it rolls into a turn to get off course. After several minutes in one direction, it reverses the roll and turns in the opposite direction.

This off-course maneuvering (aviators call it an S-turn because of its shape) increases the distance the shuttle will have to glide to the runway. The increased distance means it has more time to descend to the correct glide path without getting too hot. The beauty of this arrangement is that if, at any time, the shuttle detects a head wind or any other atmospheric phenomenon that might put it below glide path, the guidance system merely commands it to fly straight toward the field. Later, if it encounters a tail wind that's going to push it high on glide path, it resumes S-turning to lose altitude. By continually doing this—flying straight for the field or S-turning, it will arrive over the field at just the right altitude and speed to circle into landing. This procedure means that for much of the reentry, the shuttle is actually falling with one or the other wing pointed toward the earth.

### Can a shuttle commander fly a manual reentry?

Yes. A nominal reentry is done by the autopilot but the commander can take over manually at any time. If he did, however, he would have to use the shuttle's navigation displays to steer by, since looking out the window isn't going to tell him anything. Most of reentry will be over the ocean, making it impossible to know whether he's lined up on glide path and heading. Only after rolling out of the turn to final approach, which is about 1 minute before touchdown, will the commander be able to look out the window and see the runway and land the shuttle like a conventional aircraft.

### How long does reentry take?

From deorbit burn to touchdown takes approximately 1 hour.

### When does weightlessness end?

Remember, astronauts aren't weightless in orbit because of their distance from Earth, and they won't re-experience g-forces on reentry merely because they have gotten closer to Earth. Rather, weightlessness will end on reentry because the atmosphere interrupts the shuttle's freefall. Would you believe a rain of lost garbage announces this event?

There are always some small items that have floated into corners of the cockpit and escaped pre-deorbit cleanup—a lost M&M candy, a pencil, a small battery, a crumb of food. As the shuttle itself is slowed by the thin fringes of the upper atmosphere (around 70 miles high), what's to slow these loose items inside the cockpit? Nothing—at least until the floor gets in the way. So they gently drift toward the floor—a rain of lost garbage. If you want to understand reentry g-forces, imagine the astronauts are among these unrestrained pieces of debris. What would happen to them as the shuttle is slowed by the atmosphere? Just like the other items, they will continue in their free fall until the floor gets in the way. At that point, they are no longer weightless. They are being squeezed against the floor. They are experiencing g's. And the thicker the air gets, the greater the rate of slow-down on the shuttle and the more squeeze (the more g's) they will experience. The fact the astronauts are strapped to their seats doesn't change this squeeze. It's just they are being squeezed into those seats instead of against the floor.

## What are the maximum reentry g-forces?

For most of reentry the g-forces never get too high—a max of about 1.5g's. (They might briefly reach 2.5 g's in the final turn toward the runway.) However, because astronauts have been weightless for a week or two, the g-forces seem much greater. Astronauts feel as if they are being *crushed* into their seats. (On ascent, the g-forces act perpendicular to the spine, so it feels as if an invisible hand is pushing down on your chest. On reentry, the g-forces act parallel to the spine, so it feels as if the same hand is pushing down on your shoulders.)

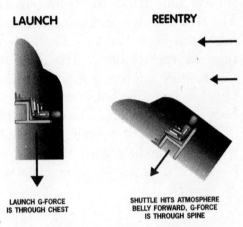

**LAUNCH**

**REENTRY**

LAUNCH G-FORCE
IS THROUGH CHEST

SHUTTLE HITS ATMOSPHERE
BELLY FORWARD, G-FORCE
IS THROUGH SPINE

*Shuttle launch and reentry g-force.*

## Does the shuttle shake on reentry?

Unlike launch, there's not much shaking aboard a reentering shuttle. The most noticeable moment of shaking comes as the shuttle's speed drops below the speed of sound. At this point, shock waves that had been trailing the shuttle are able to race ahead of it. As they move across the wings and fuselage, the shuttle shakes with a high-frequency buzz for several seconds.

## Do astronauts faint on reentry?

There have been no reports of astronauts fainting during reentry but the danger of blacking out is significant. The reason for this goes back to the orbit phenomenon of fluid shift. Remember, in the first hours of weightlessness, fluid migrated into the upper body. Besides making the astronauts' faces fatter and giving them a sensation of standing on their heads, this upward shift of fluid tricks the body into believing it is overhydrated. The result is increased urination and a loss of blood volume. In orbit, this is no big deal. The body achieves the correct hydration for weightlessness. The danger comes on reentry. Because the reentry g-forces act in a direction parallel to the spine, blood is being pulled *from* the upper body and *into* the lower body—the exact opposite of what happened in orbit. This pull of blood from the brain, combined with the fact that the weightless fluid shift has left them with lower blood volume to begin with, means astronauts face a very real threat of fainting on reentry.

### Do astronauts wear anti-g suits?

Yes. These suits are similar to what fighter pilots wear and look something like cowboy chaps. They encircle the legs and belly with air bladders. Prior to the onset of reentry g-forces, the anti-g suits are inflated with air, thus squeezing the astronauts' legs and stomachs like a large tourniquet. This constricts the downward movement of blood. With more blood in the upper body, the astronaut has a better chance of remaining conscious.

### Are there other ways to prepare the body for reentry?

Yes. In what is referred to as a fluid-loading protocol, before reentry astronauts swallow salt tablets and drink large quantities of fluids to force their bodies into a super-hydrated state. In fact, some astronauts will drink throughout reentry, holding the straw of a drink container in their mouths as if they are sucking on a pacifier. The increase in blood volume that results from fluid loading helps combat the dehydration effect of the orbit fluid shift.

### What does reentry feel like?

Because of the g-forces, a stomach-squeezing anti-g suit, and the fluid loading protocol, reentry feels as if you have a sack of cement on your shoulders, a belly-button compressed to your back bone, a fluid-bloated stomach, and a distended bladder.

### What does reentry sound like?

Reentry is essentially silent until the shuttle reaches approximately 25 miles altitude. Here, the air is finally thick enough to produce a faint rushing noise against the cockpit. This noise will slowly increase in volume until it's significantly louder than the rushing noise you hear in a cruising commercial airliner.

### What does a shuttle reentry look like?

In the early stages of reentry, the nose is tilted about 40 degrees upward, so all you see from the forward windows is the black of space. At about 50 miles' altitude, this blackness is replaced with the light show of air friction. Nothing on the shuttle melts, but the air itself becomes so super-heated it glows. As the temperature increases, the glow changes from a red to an orange and finally to a pinkish color. Also, because the shuttle hits the atmosphere belly first, air streams around the fuselage and wings and combines above the shuttle's upper windows. As it does so, its brightness intensifies. During a night-side reentry, this wake of hot air looks like a flapping ribbon of fire. Flashes of it will shine through the two upper windows like flashes of lightning.

### Does air friction make the cockpit hot on reentry?

No. However, wearing the launch/entry pressure suits does make many astronauts sweat profusely on reentry. The dehydration from this sweating can be dangerous because it increases the possibility of fainting. To eliminate this discomfort and danger, NASA is modifying the pressure suits to include a thermal undergarment that circulates cool water around the astronauts' bodies.

## What prevents the shuttle from melting on reentry?

The shuttle is made mostly of aluminum, which would melt well below the maximum reentry temperatures of nearly 3,000° F. Something has to insulate it. In the early space program, manned capsules had ablative heat shields as insulators. These carried the heat of air friction away by ablation (melting). While this was an effective method of protecting the capsule, it was a one-shot operation. The shield could not be used again. The shuttle, however, has a completely reusable heat shield. Nothing melts on reentry.

The major components of the *thermal protection system* (TPS), are the approximately 20,000 black tiles glued to the belly of the shuttle. Basically, these are made out of common sand that has been formed into very thin fibers. Since it's 90% air, the material is very light. In fact, if you could hold a tile, you would think you were holding a piece of styrofoam. Depending upon their thickness (they vary between 1 and 5 inches thick), they can withstand up to 2,300° F temperatures.

The tile material is an amazing substance. It dissipates surface heat so quickly that it can be removed from a 2,300° F oven and comfortably held by the edges in ungloved hands, even while its interior still glows red. A carbon-based shield is used on the nose and wing leading edges because these areas experience the highest temperatures on reentry—nearly 3,000° F.

## Does the heat tile design work very well?

Yes, though it did generate a lot of skepticism in the press when NASA first started flying the shuttle. The very idea of *gluing* tiles on the belly of a space ship that hits the atmosphere at nearly 18,000 mph and heats up to 3,000° F just didn't seem high-tech enough to be safe. But the shuttle TPS has proven to be very effective. On several early missions, some tiled did pop off, but the shuttle reentered with no damage. These few problem areas were modified, and now it's rare to have anything go wrong with the thermal protection system.

**Why doesn't the shuttle just do a longer deorbit burn to slow it significantly and thus eliminate the need for heat tiles?**

For a shuttle to slow down to a couple hundred miles per hour (so atmospheric friction wouldn't be a problem) would require as much energy as it took to propel it to orbit speeds in the first place. If it takes 4 million pounds of propellant to accelerate a shuttle to 17,300 mph and lift it to an altitude of 200 miles, then it will take 4 million pounds of propellant to decelerate it to zero mph at zero altitude. This is true of everything: cars, planes, bicycles, and so forth. To change an object's velocity—to go either faster or slower—always requires the expenditure of energy. The amount of that energy will be the same for identical changes in speed. In other words, if you could somehow convert into gasoline the heat energy absorbed by an automobile's brakes as it is stopped from 100 mph, you would have enough fuel to accelerate that same automobile back to 100 mph. With this knowledge, it's now clear why a shuttle can't be braked in orbit to a couple hundred miles per hour. It can't carry enough fuel to do so and, instead, must rely on atmospheric friction to do the braking.

**Why does the shuttle need a heat shield for reentry but not for ascent?**

Extreme atmospheric heating (and the requirement for a heat shield) occurs as the result of hypersonic flight in the atmosphere. The shuttle doesn't need a significant heat shield for ascent because it never experiences these conditions while it's going up. Early in ascent, it flies a nearly vertical trajectory at relatively slow speeds. This puts it quickly above the thickest part of the atmosphere. That's not to say it experiences no aerodynamic heating. There's enough friction that the tips of the SRB's have to be protected by an ablative heat shield. (The nose of the shuttle is protected by it's reentry shield; otherwise it would need some minimal protection during launch.) But the heat experienced on ascent is a fraction of what the shuttle has to take on reentry, because on ascent it's moving further and further out of the atmosphere as it accelerates to orbit speeds. On reentry, the opposite situation exists. Then, the shuttle is falling into the atmosphere at orbit speeds. It's this combination of speed and thickening air that results in tremendous friction at reentry and requires a significant heat shield.

### Does the shuttle have sensors to warn the crew of any heat tile damage before they start reentry?

No. But, if the shuttle is carrying a robot arm, it is possible to maneuver it underneath the shuttle and use the TV camera on its end to inspect the tile. I did this on my second shuttle mission. During launch, ground cameras recorded a piece of the tip of one of the solid rocket boosters (SRBs) breaking off. Mission Control feared the object could have damaged some of the heat tiles, so they directed me to use the robot arm camera to inspect the belly. When I did, we noted that several hundred tiles had been damaged. In fact, after landing we found that one tile was completely missing and 600 others were so severely damaged they later had to be replaced. In spite of this damage, however, *Atlantis's* skin (and ours) remained protected.

### Could a crew repair or replace any damaged or missing heat tiles prior to reentry?

No. Early in the shuttle program NASA experimented with several schemes to provide the crew with a heat tile repair kit, but none was ever practical. There is no way for a spacewalker to anchor herself on the shuttle belly (which would be needed to do any repairs). Also, because of the fragility of the tile, it's more likely a space-walking astronaut would do more harm than good.

### What is reentry blackout?

During the period of maximum heating, the air around any reentering spacecraft gets so hot it becomes ionized (electrically charged) and radio signals cannot be received or transmitted through it. This is referred to as blackout. In the old days, reentry blackout provided some high drama for the news reporters—"Will the crew still be alive when they come out of blackout?" Today, that drama is essentially gone. On a reentering shuttle, the ionized air blocks the belly communication antennas, but the upper fuselage antennas can usually get signals to and from Mission Control through the geosync tracking, data, and relay satellites (TDRS).

### How come people on the ground hear two booms when a shuttle is landing?

Anybody who ever watches a shuttle landing will hear two distinct booms in rapid succession about 5 minutes before touchdown. These are the sonic booms from the two strongest shock waves—the one generated by the nose and the other generated by the wings.

### Who lands the shuttle?

The commander lands the shuttle. During a nominal reentry and landing approach, the shuttle will be under autopilot control until the final 5 to 6 minutes of flight. It's then that the commander will take manual control and fly by reference to the instruments. About 1 minute before touchdown, she'll finally see the runway.

### Can the shuttle land automatically?

Yes. Though it has never been fully tested in actual flight, the shuttle has an auto-land capability that could theoretically bring the shuttle to a blind, hands-off landing. NASA, however, would probably always land the shuttle at an alternate site if the primary landing site visibility was so bad it would necessitate an instrument approach.

### Are mission specialist astronauts trained to land the shuttle?

No. It's impractical, expensive, and uselessly redundant to train all crewmembers to perform all tasks. Pilots train to fly the shuttle, while mission specialists train for payload and spacewalk activities.

**If the two pilots were incapacitated, could the mission specialists land the shuttle?**

Though shuttle pilots would be loathe to admit it, mission specialists would have no problem bringing a shuttle back without their help. All they would have to do is leave the autopilot engaged for an auto-land, lower the landing gear, pop the drag chute and step on the brakes. As we mission specialists like to joke, even a monkey or a Marine could do it.

**Why does the pilot wait until the last second to lower the landing gear?**

This is something that bothers many people. They watch on TV and think, "They've forgotten to put down the wheels!"

It's easy to understand the reason for waiting if you remember a landing shuttle is a *glider*. Let's suppose the shuttle pilot presses the gear button 15 *minutes* prior to landing instead of 15 seconds. Now, let's suppose there's a malfunction, and the wheels don't come down. What are the astronauts going to do? Remember, they have no way of leveling off and circling to try other methods of getting the gear down. The shuttle is falling like a brick. The astronauts won't be able to do anything, and that's *exactly* the same situation they would face with a gear malfunction only 15 seconds from landing. In other words, in a glider there's no difference if you have a gear-lowering problem 15 minutes or 15 seconds from landing. There's nothing you can do in either case. But there is a very good reason *not* to lower the wheels early. When they are down, drag increases, which shortens the distance a shuttle can glide. A crew might need the maximum glide distance to reach the runway. So it's better to not lower the gear until you're just 15 seconds from landing.

**How long a runway does a shuttle need to land?**

With a drag chute and maximum braking, a shuttle would use approximately 5000 feet of runway from touchdown to wheel stop. However, to save wear and tear on the brakes, normal landing stops are made with drag chutes and light braking, and the distance from touchdown to wheel stop averages about 8,500 to 10,000 feet (most commercial airliners use about 3,000 to 4,000 feet to come to a stop). The concrete runways at Kennedy Space Center and Edwards Air Force Base are 15,000 feet long.

### Why does it take astronauts so long to get out of the cockpit after they land?

Astronauts delay their exit from the shuttle for three reasons. First, there are some checklist items that need to be completed. Second, it takes the ground crew a while to use special instruments to sniff for possible leaking toxic fuel. You don't want to step out of a shuttle into an invisible cloud of poisonous gas. Finally, the crew needs time to drink more fluids and do some mild exercise to regain their Earth legs. There have been several instances of astronauts fainting in the first few moments of standing. It would hardly convey a Right Stuff image to faint on national TV while walking down the stairs.

### Why are some astronauts carried on stretchers from the shuttle?

This is sometimes done after life science research missions because the doctors want to more carefully observe—in a laboratory setting—the human body's readaptation to gravity. The crew are not being carried because they can't stand and walk on their own.

### After landing, how long does it take to get used to gravity?

For the first several hours back on Earth, you have a heavy feeling, and walking is noticeably more difficult—as if you are wearing concrete boots. Also, one's sense of balance is affected, and it's not unusual to see an astronaut trip over his own feet. (I did while walking to a microphone to address a crowd of well-wishers.) The first morning after landing provides another reminder that gravity is back. Getting out of bed requires a conscious effort because of a powerful sense of being excessively heavy.

Opinions vary as to how long it takes for these readaptation effects to disappear. Most astronauts believe the readaptation time depends upon the length of the mission. Also, some astronauts believe the readaptation time is less and less after each of their subsequent missions. In researching this book, I got answers of from 1 day to 1 week to the question of how much time it takes to get back to normal. My personal impression from my longest mission (only 6 days) was that the body was totally recovered in about 3 days.

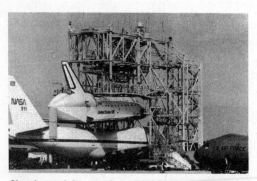

**Shuttle and its carrier aircraft in mate-demate facility.**

## How does the shuttle get on the back of the NASA 747?

If the shuttle lands at Edwards Air Force Base, it must be returned to the Kennedy Space Center via the NASA 747 carrier aircraft. It's placed on the back of this aircraft via the *mate/de-mate facility* (MDF). The shuttle is towed underneath the MDF crane, which winches the vehicle off the ground. The shuttle's wheels are then retracted. Then the 747 aircraft is towed underneath the shuttle. The spacecraft is then lowered onto the back of the aircraft and bolted into place. After flying to the Kennedy Space Center, the 747 (with the shuttle still on top) is towed into another MDF where the process is reversed—the shuttle is lifted from the back of the aircraft, the aircraft is towed away, the shuttle's wheels are lowered, and the shuttle is lowered to the ground and towed into a hangar to be readied for its next mission.

# CHAPTER 7

# Challenger

## What was the *Challenger*?

*Challenger* was the second shuttle orbiter to fly in space (*Columbia* was the first). Seventy-two seconds into its tenth mission (designated STS51-L and the 25th overall shuttle mission), it was destroyed when the right-hand solid rocket booster (SRB) experienced a burn-through in its side. The seven-person crew was killed, making them the first inflight astronaut fatalities in the history of NASA.

## Who were the astronauts killed aboard *Challenger*?

Killed aboard *Challenger* were the following individuals:

Commander: Dick Scobee
Pilot: Mike Smith
Mission specialists: Judy Resnik, El Onizuka,
    and Ron McNair
Payload specialists: Christa McAuliffe and Greg Jarvis

Four of these individuals (Scobee, Resnik, Onizuka, and McNair) were from my astronaut group and were very close friends. I had flown my rookie mission with Judy Resnik. Obviously the loss of these seven *Challenger* crewmembers overwhelmed me and the rest of NASA with profound grief.

## Why did *Challenger* blow up?

*Challenger* didn't actually *blow up* but rather was destroyed in a *structural breakup* precipitated by leaking hot gas from the side of the right-hand SRB. This gas—burning at a temperature of about 5000 degrees—impinged on the external fuel tank and the lower strut holding the right-side booster to the tank. (There are struts at the top and bottom of the fuel tank that hold the boosters to it.) At about 72 seconds into flight this strut gave way, essentially detaching the bottom of the booster from the fuel tank. As the bottom moved outward from the tank, the top of the booster rotated into the fuel tank causing it to rupture and starting the disintegration of the entire vehicle. (For the purposes of this book, I will continue to refer to *Challenger*, as everyone else does, as an *explosion*.)

**What caused the booster rocket to leak?**

To answer this question, it is first necessary to understand how the booster rockets are assembled. Each booster rocket is 150 feet long and weighs nearly 1.3 million pounds, making it impossible to ship them as one

*Fire leaking from right-hand solid booster during Challenger ascent.*

piece from Utah (where the factory is located) to Florida. So, the rockets are built in four propellant-filled pieces that are transported separately by train to the Kennedy Space Center.

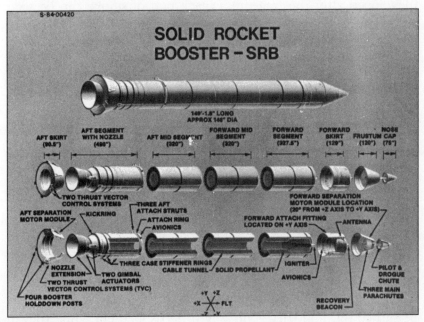

*Booster rocket components.*

Once there, they are stacked on top of each other and bolted together. A pair of rubber O-rings inside the joint are intended to prevent the fire from leaking out. (They are called O-rings because they are in the shape of a O, i.e., a circle.) How, you might ask, can rubber withstand 5000 degrees of heat? It can't, but it doesn't need to. Several inches of insulating material separate the O-ring from the combustion fire, and, by sealing the joint, the O-rings prevent any flow of gas toward them. If the seal works properly, the gas has no place to go except out the bottom of the rocket, and the rubber stays well within temperature specifications. The safe operation of this design hinges on the ability of the O-ring to form a gas-tight seal. That's why NASA put *two* O-rings in the joint. As long as at least one works in each joint, there will be no leaks. But, if both ever leak, the hot gas will be free to travel toward the rubber, melt it, and begin to melt the steel casing. This is exactly what happened on *Challenger*. It is theorized the very cold temperatures at the time of *Challenger's* launch (about 36°) caused the rubber O-rings to become stiff and lose their ability to seal. In launch movie film (developed after the tragedy) a black puff of smoke can be seen at the lower segment of the right booster. This marks the failure of both O-rings.

### Did the *Challenger* crew know they had a leak in their booster rocket?

No. Nobody knew there was a leak. The crew didn't know and neither did Mission Control. There are no sensors aboard the boosters to warn of a leak. NASA photos showing the leak are from motion picture film that was developed many hours after the tragedy.

### If the crew or Mission Control had known of the leak, could the astronauts have jettisoned the boosters?

No. Unlike the liquid-fueled engines, the booster rockets cannot be turned off, and there's no way to jettison them while they're still burning. There is an SRB separation button in the cockpit, but it's there to serve as an emergency backup to the automatic jettison of the *burned out* boosters. Nobody can be certain what would happen if it were pressed while the boosters were still thrusting, but most engineers believe the enormous force of their thrust, pushing them into the attachment struts, would hold them on the tank.

### If the crew or Mission Control had known of the leak, could the astronauts have separated the shuttle from the fuel tank?

Yes. There is an external tank separation button (ET SEP button) in the cockpit that, if pressed, would separate the shuttle from the tank. Since the boosters are attached to the tank, this would get the shuttle away from them. Unfortunately, most engineers believe if the ET SEP button was ever used while the boosters were still thrusting, the shuttle would be peeled off the tank and immediately flipped out of control. It would spin into the ocean and the astronauts would still be killed. The bottom line is that the crew has no options to safely get away from a failing booster. The SRBs *must* work.

### Did the *Challenger* crew survive the explosion?

Several weeks after the disaster, the cockpit—containing the remains of the crew—was found by salvage divers about 25 miles offshore in about 85 feet of water. Autopsies were performed but were unable to fix the time and cause of death. From engineering analysis it was definitely determined that the g-forces the crew experienced at breakup would not have been fatal or even incapacitating. This fact, and another piece of evidence, suggests that death occurred at water impact. The additional evidence was in the form of a switch on an emergency air-breathing container attached to the back of pilot Mike Smith's seat. This switch was found to be in the on position. In a nominal mission this switch remains off, implying that whoever turned it on was doing so *after* the explosion. (The vehicle breakup would have cut off oxygen

*The Challenger is destroyed by a leaking solid rocket booster.*

to the crew helmets and made it necessary to either activate this switch or raise the helmet face plate to continue breathing.) Could the switch have been moved by the crash itself? NASA examined this possibility and felt that it was unlikely the crash would have caused the switch to flip. It seems most likely the switch was moved by El Onizuka, who was sitting behind the pilot. These facts suggest only that the crew might have *survived* the explosion. They do not imply the crew was *conscious* to water impact.

## Was the crew conscious during the cockpit's fall to the water?

Again, it's impossible to positively answer this question. When films of the disaster were carefully analyzed, the cockpit was seen tumbling away from the explosion. It had wires and tubes trailing from it, but otherwise it appeared intact. To know if the crew was conscious at this point requires knowledge of the pressure integrity of the cockpit. The explosion occurred at 46,000 feet and the shuttle's upward momentum carried the separated cockpit to a peak altitude of about 60,000 feet. At these altitudes the ambient air is too thin to keep a person conscious. Therefore, if the cockpit lost cabin pressurization, the crew would have passed out and remained unconscious to water impact (let's pray this was the case). An explosive decompression (e.g., from a window breakage) would have brought almost instantaneous unconsciousness (within 3 seconds). Unfortunately, the fact that someone had time to flip Mike Smith's emergency air switch implies such an explosive decompression did not occur. But, even if the leak weren't explosive, it might have been rapid enough to bring unconsciousness within 10 to 15 seconds. Since it is impossible to know the state of the cockpit's pressure integrity in its fall, it's impossible to know when or if the crew blacked out.

### Is it possible the crew survived water impact and drowned before rescue forces could get to them?

No. If the crew was alive in the fall, death was instantaneous at water impact.

After a 2-minute 43-second fall (measured from the instant of the explosion), the cockpit slammed into the sea surface at over 200 mph. Such an impact is no more survivable than driving a car into a brick wall at the same speed.

### Why wouldn't their emergency breathing air have kept the *Challenger* crew conscious?

At the peak altitude reached by the cockpit debris (around 60,000 feet) a person needs *pressurized oxygen* to breathe to stay conscious. The emergency air the *Challenger* astronauts carried was just that—*air*, not oxygen. It was never intended for inflight use. Instead it was designed to protect a crew from toxic fumes in the event of an emergency ground escape. It can only be assumed that a *Challenger* crewmember's reaction to turn on this air supply was a survival response to a deadly situation that had no other response alternatives.

### Why didn't their pressure suits keep the *Challenger* crew conscious?

The *Challenger* crew wasn't wearing pressure suits. Beginning with the operational era of the shuttle—the fifth shuttle flight—crews launched in coveralls. There is some irony in the fact the *Challenger* crew was not protected by pressure suits. Without this protection—and if there were a rapid decompression (leak) in

the cockpit after the explosion—a mercifully quick unconsciousness would have resulted.

### Could the crew have bailed out?

No. Even if they were alive and conscious after the explosion,
the *Challenger* crew had no parachutes or ejection seats. Beginning with the fifth flight of the shuttle and through *Challenger*
(*Challenger* was the 25th flight), shuttle crews had no bailout system of any kind. The first four shuttle flights did have two ejection seats to give the two pilots a way to bail out, but those were
deactivated after the fourth flight. The reason for this goes back
to the very beginning of the shuttle design. Then, NASA felt
they could build a rocket so safe that the crew would never need
to bail out. Still, nobody wanted to risk people's lives unneces-
sarily in the first few test flights. So the shuttle design was made
to include two ejections seats (and only two) to protect the two-
person crew that would fly on these first flights. After four very
successful test flights, NASA believed the shuttle design had
been proven. The shuttle test phase was declared over, the ejection seats were deactivated (and later removed), and the fifth
shuttle flight marked the first operational mission.

### Are there recordings of the *Challenger* crew talking to each other before and after the explosion?

Stories of a NASA coverup of a postexplosion crew voice recording persist in the press. But these stories are false. No such record-
ings exist. The shuttle cockpit does contain a tape recorder that
records the crew's intercom comments, and this recorder was recovered from *Challenger's* wreckage. Investigators were able to re-
play the crew conversations but only up to the moment of the
explosion. The tape reveals the crew had absolutely no idea the
booster rocket was failing. The voices are completely normal
through the final comment, when the words "uh oh" can be heard
an instant before the explosion. It sounds like pilot Mike Smith's
voice, but it's not a scream or a shout. It's said in a normal tone

and could have been in reference to something totally unrelated to the booster malfunction (e.g., "Uh oh, my checklist came loose.") Then, again, he might have been preparing to comment on a small guidance error due to the leaking gases. It's impossible to know.

But it is a *certainty* that the recording ends at this point and NASA is not hiding postexplosion conversations of a dying crew. Such a recording was an impossibility because the cockpit (where the recorder was located) was instantaneously ripped from the rest of the shuttle's fuselage by the explosion. Since the only source of electrical power in the shuttle is from fuel cells located under the cargo bay of the shuttle, this violent fragmentation literally pulled the plug on electricity to the cockpit and stopped the voice recorder and all other electrical equipment. Nothing could have been recorded after that.

### Do shuttle astronauts now have a bailout system?

Yes, but it's very primitive and (in my opinion) virtually useless. NASA desperately wanted to retrofit the shuttle with some type of automatic bailout system to protect the crew in the event of another launch emergency. Engineers looked at various schemes for ejection seats, extraction rocket packs, and a cockpit escape pod. Unfortunately there was no practical way to make such design changes. To explosively remove crewmembers safely out of the cockpit, like an ejecting fighter pilot, would require huge, rocket-propelled seats and explosive overhead hatches. An escape pod would require explosive pyrotechnics to sever the entire cockpit from the fuselage, a rocket to blast it away from an exploding shuttle, room for a large parachute to lower it to the ground, and a floatation device. After much study, all of these designs were deemed unsafe and/or impractical.

So, NASA engineers gave the crew the only escape system that was practical—a backpack-worn parachute. Basically, post-*Challenger* shuttle crews fly with an escape system similar to a World War II bomber bailout system. Then, crews had to unstrap from their seats, get to a hatch, and jump out. Current-day shuttle crews would use identical procedures: unstrap, walk to the side

hatch, and jump overboard. There is one modification to our bailout system that distinguishes it from a World War II bomber bailout. If astronauts jumped out of the shuttle side hatch, they would hit the wing and be killed. To prevent this, they have a pole to slide down and get under the wing. The pole is in a housing inside the cockpit. If a bailout is required, a crewmember will jettison the side hatch and release the pole (springs push it out and downward). The astronauts then clip their harnesses to rings that encircle the pole. These rings enable them to slide down the pole and under the wing. Upon separation from the pole, the parachute deployment system is automatically activated.

*A test parachutist evaluating bailout system in an Air Force aircraft.*

### How long would it take a shuttle crew to bail out?

In ground tests, it takes about 1 minute for a seven-person crew to bailout. In a real emergency, this means that the first crewmember would be out of the hatch at around 25,000 feet and the last (the commander) would be jumping overboard at around 12,000 feet. Bailout can only occur from a shuttle that is in a controlled glide, so the last crewmember to bail out will always be jumping from a lower altitude. In a condition of a controlled glide, it's likely that every crewmember would be able to slide down the pole and successfully parachute away.

### Why doesn't NASA just reinstall the two ejection seats that were on the first four shuttle missions?

This could be done, but remember, the shuttle design only permits *two* ejection seats—for the commander and the pilot. Limiting the shuttle crew size to two people would eliminate almost all shuttle missions. It needs a minimum of five people to do meaningful work. The other alternative—to fly five people with only two ejection seats—is morally indefensible. How would you like to take a commercial airline flight knowing that only the two pilots could bail out in the event of an emergency? If you can't give *everybody* a way to bail out, then you shouldn't give *anybody* a way to bail out. To do otherwise is to imply some lives are more important than others.

### If the new bailout system had been aboard *Challenger*, could it have saved the crew?

In my opinion, the answer to this question is *no*. The backpack bailout design is an extremely primitive system. It will not protect astronauts in the scenarios that most typically require bailout from a high-performance craft—while the engines are running and/or the vehicle is out of control. In other words, bailing out using a backpack parachute is only going to work if the shuttle's engines are off and it's in a controlled glide. This is the only condition in which an astronaut can get out of her seat, walk to the side hatch (or climb down the stairs if she's in the upper cockpit), deploy the pole, attach her harness to a ring, and slide out. In conditions in which the engines are thrusting or the cockpit is tumbling, the g-forces will pin the crew to their seats. I believe the latter situation applies to the *Challenger* crew. If anybody had survived the explosion and was conscious, they wouldn't have been able to leave their seats and make it to the hatch to bail out.

Other astronauts disagree with my answer to this question. Some believe the *Challenger* cockpit was stabilized in its fall and one or two crewmembers might have been able to crawl to the side hatch and bail out with the backpack parachute (as we are now carrying).

### If backpack parachutes have such a low probability of protecting shuttle astronauts, why do they wear them?

First, the bailout system is relatively easy to carry. Second, there may be some strange, one-in-a-million failure that results in the crew being in a controlled, gliding shuttle that isn't going to make it to a runway. Since an ocean ditching or a crash landing would almost certainly be fatal, the bailout system could be used to save the crew.

## Would you ever fly the shuttle knowing it doesn't have a very effective escape system?

Yes, and I did. I flew twice after *Challenger* with the above-described bailout system, and if I were still at NASA, I wouldn't hesitate to take more flights. But I would do so with full knowledge that I had no effective escape system. I suspect most other astronauts feel similarly. It's the NASA team doing a quality job that's going to protect them, not a parachute.

## What changes did NASA make to the shuttle booster design after *Challenger*?

During the 33 months the shuttle was grounded, the booster rocket segment joints were redesigned. They were made much stiffer, a third O-ring was added, and heaters were installed to keep the O-ring rubber warm on cold days.

## What changes did NASA make in the operation of the shuttle after *Challenger*?

To answer this question, it's first necessary to understand the satellite launch market and how the shuttle was intended to compete in that market. The vast majority of satellites can be divided into three categories: commercial, military, and research. Prior to the shuttle, all of these satellite types were being launched by unmanned, expendable rockets (Delta, Atlas, Titan, and the French Ariane rockets). The original intent of NASA was to use the shuttle to compete against these other rockets, and prior to *Challenger*, it did just that. The shuttle was used to launch many commercial, military, and research satellites.

After *Challenger*, however, NASA had to cut the launch rate to safely fit its human and hardware resources. This meant they could launch only approximately seven or eight missions a year—far too few to compete in all satellite markets. So, it curtailed its commercial and military shuttle operations. Except for a few military missions that were too far along in construction to switch back to the older rockets, the shuttle has been used, almost exclusively, as a science spacecraft since the *Challenger* tragedy. It does not launch any major military payloads or commercial communications satellites.

### Where is the *Challenger* wreckage?

It was placed in an abandoned missile silo on Cape Canaveral Air Force Station.

### Where is the *Challenger* crew buried?

Identified remains were returned to the families for interment. The unidentified remains were cremated and interred in a common grave in Arlington Cemetery.

### Did the nine missions that *Challenger* flew prior to the explosion pose a greater risk to astronauts than those riding in other shuttles?

I am often asked if I flew the orbiter *Challenger*. When I reply, "No," people typically respond, "You were lucky." The obvious implication of this response is that the shuttle *Challenger* was somehow more dangerous than the other shuttles. This is not true. Remember, *Challenger*—the orbiter—didn't blow up. The tragedy was caused by a flaw in the booster rocket design. Every mission prior to *Challenger's* final mission (there were 24 of them, using the orbiters *Columbia*, *Discovery*, *Atlantis* and *Challenger*) flew with the identical flawed booster design.

# CHAPTER 8

---

# Astronaut Facts

### When did you first want to be an astronaut, and how did you become one?

As far back as I can remember—even as a very young child in the early 1950s—I had a passionate interest in the sky. My dad was in the Air Force and would take my brothers and me to his base and let us sit in airplanes, so perhaps this exposure planted the seed of being a pilot. It's really impossible to know. For whatever reason, though, I longed to be a jet fighter pilot. In 1957, with the launch of *Sputnik* and the dawn of the space age, that dream changed. I wanted to be an astronaut. I was 12 years old.

By the time I graduated from high school in 1963, the biographies of the early astronauts that I had read convinced me that I had to become a military test pilot to pursue my astronaut dream. This led me, by a circuitous path, into the Air Force. I wanted to attend the Air Force Academy but my Scholastic Aptitude Test scores were too low, so I entered West Point instead. (Apparently, West Point felt it didn't take a rocket scientist to lead an infantry platoon.) In 1967, at my graduation from the Military Academy I took a commission in the Air Force with

Author as teenager in the New Mexico desert with a homemade rocket. *(From author's collection)*

the intention of entering pilot training. Unfortunately, I was unable to meet the eyesight requirements and ended up as a Weapon Systems Operator flying in the backseat of F-4 Phantom jets (I was a "Goose," for all of those familiar with the movie *Top Gun*). I was certain my dream of being an astronaut was over, since astronauts of that era had to be pilots.

During the following 10 years of my Air Force career I did two things that ultimately proved critical in becoming an astronaut. First, in 1975, I completed a master's degree program in aeronautical engineering at the Air Force Institute of Technology. Next, in 1976, I graduated from the Flight Test Engineer course of the USAF Test Pilot School. I was still unqualified for astronaut because I wasn't a pilot. Then, a year later, the rules changed. In NASA's 1977 announcement to select astronauts for the space shuttle program, a new crew position appeared—Mission Specialist (MS) Astronaut. Mission specialists were to be the people who operated the shuttle robot arm to release and capture satellites, to do spacewalks, and conduct experiments, and they didn't have to be pilots! I immediately submitted an application and was selected in the first group of shuttle astronauts in 1978.

**How does NASA select new astronauts?**

NASA never solicits individuals for the astronaut program. If you want in, you must apply. NASA will accept applications at any time from civilians. Military personnel must apply through their parent services, which periodically convene boards to select names to be forwarded to NASA for consideration. Approximately every 2 years, NASA will assess past astronaut retirements and future astronaut requirements and select new astronauts from this pool of applications. The NASA selection board will consist of senior NASA managers and astronauts who will review the resumes of applicants—civilian and military—and will select a group to come to the Johnson Space Center for physical exams and face-to-face interviews. The number of interviewees will be a function of how many astronauts NASA intends to pick. Usually they will call five to six times as many people for interviews as they plan to select as astronauts.

**What is the astronaut candidate interview like?**

The selection committee interview is an informal, 1-hour, free-ranging discussion in which the committee members try to measure the individual's suitability for the astronaut program. They are particularly interested in determining if the applicant is team oriented. By the very fact that the individual has been called for an interview, their academic credentials have already been accepted, so the board asks very few, if any, technical questions. More than anything else, NASA, like every business, is looking for employees who will work well with other team members.

Individuals who project an image of aloofness and superiority, as did some of the pioneering solo astronauts, will probably not make the cut. An arrogant, loner test pilot flying in a one-person capsule might have served the purpose back in the early days of the space program, but now space missions are complicated team efforts involving people of all stripes—scientists, military flyers, men and women, white and minority, even foreign nationals. NASA wants people who will excel on that team and so the board will let the interview wander into areas that might reveal the applicant's team disposition—like the person's hobbies and interests, their likes and dislikes of past jobs, their reasons for wanting to be an astronaut. It's an inexact science to intimately measure any stranger in a 1-hour interview (there are other, informal group get-togethers where the candidates are also being observed), but NASA does its best and it's been my impression that they do a pretty good job. The vast majority of astronauts I have worked with have been superb team players.

The committee is also careful to explain to candidates what the astronaut job isn't—being a researcher on the cutting edge of science. Many people are under the misimpression that astronauts invent things or discover things. We do not. We are systems operators—glorified bus drivers, really. While we might help determine where an experiment switch should be located on the instrument panel, we don't really participate in the design of the experiment or the analysis of data. We certainly didn't design the space shuttle or any of its systems. The astronaut selection committee wants prospective astronauts, particularly young, eager scientists, to understand that this lack of substantive research might make the job unsuitable for their personal goals.

An interesting fact about the astronaut interview process is that the board asks all candidates to write a one-page essay on "Why I Want to Be an Astronaut." Some people have written serious answers (I did), while others have composed humorous essays. Here's the text of my essay:

November, 1977

Flying was the first love of my life and as early as possible I did what I could to satisfy that love. As a high school student, I took flying lessons and had about 20 hours of solo time prior to graduation.

In addition to flying, I had developed an intense interest in the space program, rocketry, and the engineering aspects associated with atmospheric and space flight even before *Sputnik* I was launched. Again, I did whatever I could to satisfy these interests. While a high school student, I designed, constructed, and launched a great number of solid propellant rockets as part of my personal *Apollo* program, the objective of which was to develop a recoverable capsule. I still rate these experiments as the most gratifying accomplishments of my life. The first time a recovery parachute opened over my capsule (an empty coffee can) and lowered it successfully to the ground, I sent up a cheer that would rival anything ever heard at Houston Mission Control.

Throughout my adult years, my love of flying, engineering, and rocketry have continued to grow, and I have aimed the direction of my life toward the fulfillment of these "loves." It was a bitter pill to swallow when told I could not be a pilot because of my visual acuity (or rather lack of it). But, with that single exception, I have directed my career to satisfy my life-long interests in flight. My assignment to navigator training, tactical fighter operations, graduate school, and test pilot school were deliberate results of my efforts toward self-fulfillment and not accidents of the Air Force personnel system.

Now, employment is available in a program in which flight, rocketry, and engineering have been married into a single job specialty. It is only logical that I should seek that employment as further fulfillment of my personal interests. My answer then to the question of "Why do I want to be an astronaut?" is a selfish one. I seek that job because it will be an additional step toward satisfying a deep, intense, personal love of flying and machines that fly. But I also firmly believe that in the process of this self-fulfillment I will further the aims of NASA and the US manned spaceflight program.

### Do you have to be an American citizen to be a NASA astronaut?

Yes. But citizens of other countries have flown aboard the shuttle as mission specialist and payload specialist astronauts. No foreign nationals have flown as pilot astronauts. The European, Canadian, Japanese, and Russian space agencies all have their own astronaut programs and select astronauts to fly on joint missions with NASA.

### What physical requirements must be met to be an astronaut?

The physical exam will be the equivalent of a FAA class I exam (for pilot) or class II exam (for mission specialist), though it will also include a proctology examination, a treadmill stress test, an EEG brain wave test, and a psychiatric interview. On average, approximately 15 % of candidates fail the physical. The exam has none of the bizarre stress testing that is depicted in the movie *The Right Stuff*.

### Can you wear glasses and still be an astronaut?

Yes. NASA requires that pilot astronauts have uncorrected distant visual acuity of 20/50 or better, correctable to 20/20. Mission specialist candidates must have uncorrected distant visual acuity of 20/200 or better, correctable to 20/20.

### Can you be color blind and still be an astronaut?

No. There are too many displays and controls that require color vision to correctly interpret and operate to allow color blind individuals to be astronauts.

### Can you have high blood pressure and be an astronaut?

To be selected as astronauts, pilot and MS candidates must have blood pressure lower than 140/90 measured in a sitting position. After astronaut selection, individuals with high blood pressure can still fly space missions as long as that pressure can be controlled with medication to the 140/90 level.

### If you have asthma, can you be an astronaut?

Asthma that requires regular treatment is disqualifying. A childhood diagnosis of asthma is not necessarily disqualifying and would be evaluated on a case-by-case basis.

### Can you be an astronaut with fillings in your teeth?

Yes. Weightlessness has no effect on teeth fillings. Before a mission, however, astronauts see a NASA dentist to make sure they don't have a loose filling or some other problem with their teeth that might cause a problem in orbit. The last thing you want to have in space is a toothache. But, if a filling should pop out during a mission, our emergency medical kit does have material to make a temporary replacement filling. Also, if there is significant pain associated with a broken filling or some other dental problem, the medical kit also comes with Novocaine. If you need an injection of this pain killer, though, there are currently no dentist astronauts, and there may not even be a medical doctor on board to give you the shot. In that case, a fellow crewmember would play dentist and do it. Good luck.

### Could physically impaired people be astronauts?

Not in the foreseeable future. However, some day, when transportation into and out of orbit is more like commercial aircraft flying—very safe and reliable—I believe people who have lost the use of their legs through disease or accident will live and work in space like anybody else. I say this because you don't need legs to operate a space experiment or laboratory. On Earth we need legs for locomotion. In weightlessness you use your hands to move around. In fact, the only thing legs are good for in weightlessness are as anchors. (Foot restraints are used to anchor the body so both hands are free to work.) For leg-disabled individuals it would be trivial to design body restraints to serve the same purpose.

Another future medical possibility for spaceflight is in the treatment of burn victims. On earth, some part of the body is always in contact with a bed, making the treatment and healing of burned tissue more difficult. Weightlessness will solve that problem.

### What are the height requirements for astronauts?

To fly aboard the space shuttle, an MS astronaut must be between 4 feet 10.5 inches and 6 feet 4 inches tall. Shuttle pilots must be between 5 feet 4 inches and 6 feet 4 inches tall.

### What are the weight restrictions on shuttle astronauts?

There are none. There are female astronauts who weigh around 100 pounds and male astronauts who tip the scales at around 220 pounds.

### Do you have to be in the military to be an astronaut?

No. About 39% of shuttle astronauts have been scientists and engineers with no military background. Some of these have been university employees doing post-doctoral research. Others have come from hospital internships. Still others have been in various aerospace industry corporations.

### Do military astronauts leave their services when selected as astronauts?

No. Military astronauts are on loan to NASA. They remain on active duty and are paid and promoted by their parent service. Being an astronaut has no significant effect—good or bad—on a person's military career.

### Can military astronauts be forced to return to their parent service?

In the military there are no guarantees, so, theoretically, a parent service could demand that one or more of their astronauts be returned from NASA. However, this has never occurred. Several astronauts have returned to their parent service but only because they wanted to.

### Do you have to be a pilot to be an astronaut?

No, except that you must be a pilot if you want to be a pilot astronaut. Remember, there are two types of astronauts, pilot astronauts and MS astronauts. Mission specialist astronauts are not required to be pilots because they don't fly the shuttle.

### Do you have to be a test pilot to be an astronaut?

Theoretically, the answer to this question is no. NASA only requires that pilot candidates have at least 1,000 hours of pilot-in-command time in jet aircraft. But the application brochure goes on to say that "flight test experience is highly desirable," and so far, 100% of NASA's pilot astronauts are graduates of military test pilot schools.

### Can civilian pilots be pilot astronauts?

Theoretically, a civilian airline pilot with 1,000 hours of jet time and the proper academic background could apply to be a pilot astronaut. Practically, however, their chances of selection are dim. As of the astronaut class of 1995, every pilot astronaut ever selected for the shuttle program has had a military test pilot background. Sure, some pilot astronauts are *civilians*, but if you look deeper into their biographies, you will find they all had their start in military aviation (they are former military test pilots). The fact is if you want to be a shuttle pilot, join the Air Force, Navy, or Marines and get into their pilot training programs. While several US Army and Coast Guard pilots have been selected as MS astronauts, none has ever been selected as a pilot astronaut.

## What type of education do you need to be an astronaut?

The minimum education requirement for both pilot and MS astronauts is a bachelor's degree in math, science, or engineering. Among pilot astronauts 60% exceed this minimum education (with master's or doctoral degrees), while 95% of the MSs exceed it. In fact, 55% of MSs (and most of these are civilian MSs) have doctoral degrees. These statistics clearly indicate that an individual competing for an MS astronaut position will not be competitive without an advanced degree, and if you are competing for a civilian MS position, you really should have a doctorate. Pilots can be competitive with just a bachelor's degree because they are being hired for their flying skills.

## Does a degree from a certain university or from a military academy increase the chances of being picked as an astronaut?

No. An applicant's degree need only be from an accredited university. The snob factor of the institution won't improve your chances for selection. Purdue University advertises itself as a leader in astronaut alumni, but this is a somewhat misleading claim. The Air Force sends many of its officers seeking graduate engineering degrees to Purdue and some of these have gone on to become astronauts, thus the high percentage of Purdue graduates. But this statistic refers to *graduate* degrees. If you look at where astronauts did their *undergraduate* studies, you'll find a wide representation of institutions—private and state, large and small, famous and obscure. Also, there are a lot of military academy graduates (West Point, Air Force Academy, and Naval Academy) represented among the military astronauts, but this is because many test pilot school graduates are academy graduates, and NASA (so far) has only hired military test pilots as pilot astronauts.

**Will attending Space Camp improve a person's chances for astronaut selection?**

Nowhere on NASA's astronaut application does it ask if the applicant has attended Space Camp, so the answer to this question is no. The application really focuses on the individual's college performance and subsequent job history. Childhood activities aren't queried (though they might come out in the face-to-face interview). That's not to say, however, that a Space Camp visit, Boy or Girl Scouting program, or other childhood experience wouldn't be important—indeed, wouldn't be the *key*—to a person's successful pursuit of the astronaut dream. One of these activities might just plant the seed for that pursuit. Certainly my dream to be an astronaut began in my childhood.

**Is there anything else a person could do to increase his or her chances of being selected as an astronaut?**

My answer to this question is just a personal opinion, but I believe an applicant with a fluency in the Russian language would be significantly competitive, even if they only met the minimum education requirements. The astronaut application brochure makes no mention of language fluency. But considering that the Russians will be virtual co-partners with NASA in the construction and operation of the international space station, it doesn't take a rocket scientist to see Russian language fluency as a tremendous résumé *enhancer*. As it is now, the astronaut office has a difficult time getting astronauts to volunteer for training with the Russians because of the language barrier. It's also a drain on the astronaut office to have to send somebody to language school for several months to acquire a minimal fluency.

**How old do you have to be to be selected as an astronaut?**

No age limit is specified in NASA's application literature. However, to have the flying background (pilot astronaut) or academic background (MS astronaut) to really be competitive for selection makes the average candidate about 34 years old. Tony England was the youngest astronaut ever selected, at age 25. John Phillips was the oldest at age 45.

**How many new American astronauts are normally picked?**

Not many. Here are the figures for all shuttle-era selections:

| Year | No. of American Astronauts Selected |
|------|------------------|
| 1978 | 35 |
| 1980 | 19 |
| 1984 | 17 |
| 1985 | 13 |
| 1987 | 15 |
| 1990 | 23 |
| 1992 | 19 |
| 1995 | 19 |
| 1996 | 35 |
| TOTAL | 195 |

In addition to these 195 American astronauts, 19 foreign nationals have also been selected as MS astronauts.

**How many men have been selected as shuttle astronauts?**

Of the 195 total American astronauts selected in the shuttle era, 159 have been men.

**How many women have been selected as shuttle astronauts?**

Of the 195 total astronauts selected in the shuttle era, 36 have been women. The first women (in fact, six women) were selected in my 1978 astronaut class.

## How many minority men and women have been selected as shuttle astronauts?

Of the 195 total American astronauts selected in the shuttle era, 26 have been from minority groups, including African Americans, Asian Americans, Hispanic Americans, and one Native American. Guy Bluford was the first African-American man in space. El Onizuka was the first Asian American (he was killed aboard *Challenger*). Franklin Chang-Diaz was the first Hispanic man. Mae Jemison was the first African-American woman in space. John Herrington is the first Native American selected as an astronaut (1996) and has not yet flown in space.

## Where does someone get an application package for the astronaut program?

Write to

NASA, Johnson Space Center
Astronaut Selection Office
ATTN: AHX
Houston, TX 77058

## Is there a mandatory retirement age for astronauts?

No. As long as an astronaut can pass a physical, she can still be assigned to a crew. But many astronauts over 50 years old are moved into management positions and fly desks for the rest of their careers.

## How are mission crews selected?

The Chief of the Astronauts and his boss, the Director of Flight Crew Operations, work together to make crew selections. The final assignments are then passed to the Director of the Johnson Space Center and the NASA Administrator for final approval.

This is a change from my era. Then, the entire process was cloaked in secrecy, but our impression was that *all* crew assignments were done exclusively by the Director of Flight Crew Operations. His decisions seemed absolute and final.

### Who determines how often you fly?

Astronauts have virtually no say in their mission assignments or their frequency of flights. These decisions are made by the Chief of the Astronauts and his boss, the Director of Flight Crew Operations.

### Can an astronaut turn down a flight assignment?

Yes, but it would almost certainly be detrimental to the astronaut's career. Because of the secrecy surrounding flight assignments, I can't positively state whether anyone has ever turned down a mission, but when I was an astronaut, rumors persisted that one individual had done so and was never again offered another flight.

### Can an astronaut request a specific flight?

Sure, and you'd get it, when pigs fly in space. Why astronauts have no say in flight assignments remains a mystery, but it is probably a tradition-based facet of NASA bureaucracy.

### Is individual compatibility a criteria for crew selection?

No. Many people are surprised to hear this. After all, they reason, being cooped up in a small cockpit with somebody you can't stand wouldn't be a lot of fun. Mission effectiveness may even be jeopardized. But astronauts are professionals, and most are able to swallow back their dislikes in the interest of the mission. Also, shuttle missions are relatively short—a week or two at most. Most astronauts could fly with the devil himself for a period that brief. But living for months on a space station or spending years together on a Mars trip will be something entirely different. Not only will there be the normal personality conflicts on these missions, but the additional problems of male–female relationships (including sexual attraction and jealousy) will significantly complicate crew compatibility. NASA understands this and has begun to investigate human relationships among astronauts to determine how best to deal with that old burden—our humanity.

### Do all astronauts fly in space?

Yes. NASA doesn't hire extra astronauts who might not be needed for space missions. Every astronaut is hired to fly into space. Of course, somebody could be medically grounded and never fly, though this has never happened. (A handful of astronauts have been medically grounded for significant periods—for example, Deke Slayton, an Original Seven astronaut who suffered from a heart murmur. Eventually, he and the few others were returned to flight status and went on to fly at least one space mission.) Also, there have been rare occasions when new astronauts have decided on their own to quit NASA prior to spaceflight.

### How long do you train for a mission?

Astronauts are normally assigned to a specific mission about 1 year prior to the planned launch date and begin their mission-specific training at that time. For exceptionally complex missions, a Hubble Space Telescope repair or a *Mir* rendezvous mission, for example, crew assignments will be made as much as 2 years prior to flight. Training for a typical mission will require about 1300 hours for MSs and 1,500 hours for pilots.

### What do astronauts do between missions?

After returning from a mission, astronauts are assigned to support jobs. For example, an astronaut might be assigned as a Capcom in Mission Control or help test new software or oversee the astronaut interface with a new payload. Whatever they are assigned to do, it's still training of sorts. Astronauts never stop training.

## Are there backup crews in training?

One of the infamous moments in astronaut history was the removal of Ken Mattingly from the *Apollo* 13 crew just days prior to launch. Flight surgeons thought he had been exposed to measles and were afraid he might get sick on the way to the moon. The astronaut assigned as his backup—Jack Swigert—replaced him on the mission. Had someone not been trained to step into his place, the mission would have been delayed while Mattingly was observed for signs of illness (which would never have come, because he never did get sick).

The NASA policy of assigning backup crewmembers to each mission continued through the early days of the space shuttle program. As the shuttle mission rate expanded, however, it became impractical to have a backup crew on every mission. Also, with a large number of flight-experienced personnel in the astronaut office, there were plenty of astronauts available to step into a mission on short notice. So, except for special missions, backup crew assignments were ended with the fifth flight of the shuttle. The exceptions occur when there are highly visible, time-critical missions that require specialized training. For example, there were backup crew members in training for the Hubble Space Telescope repair mission. This mission required intense spacewalk training, and NASA couldn't take the risk of a lengthy launch delay because of a primary crewmember illness or injury right before the mission.

It's a popular joke in the astronaut office that backup crewmembers would like to buy their counterparts a ski vacation to the Colorado mountains. The obvious aim of such a thoughtful action would be to see the primary astronaut break a leg so the backup could take the glory mission. Also, it's joked that a primary crewmember should never walk down the stairs in front of her backup for fear she will be shoved into an injury.

## What would happen if an astronaut got sick right before launch?

If a mission was so critical it *had* to launch without significant delay, NASA would replace the sick or injured crewmember in question and launch as quickly as possible. Most missions, however, aren't that time critical, so a delay for a week or two for somebody to recover from the flu (for example) would probably occur. This actually happened on my last shuttle flight. Our commander, John Creighton, came down with an upper respiratory infection while we were at the Cape waiting to launch. The flight was delayed a couple days until he got better. Of course if it had been a broken leg or something that would have taken weeks or months to mend, the mission would have gone as soon as a replacement could be trained. Probably, NASA would have just picked somebody who had recently landed from a similar mission as the replacement.

The shuttle pilot is probably the most replaceable of any crewmember on any flight since the responsibilities of that position are almost identical from mission to mission. The MS joke is that the pilots are hardly more than trained monkeys. They just watch lights and flip switches hoping for the banana pill to pop out. That's an exaggeration, of course, but their duties usually have little to do with the mission payload. They oversee the shuttle systems, and the procedures for that oversight don't change much from mission to mission. Knowing their replaceability, I doubt any pilot, on the eve of a mission, would voluntarily tell the flight surgeon he was feeling ill and some pilots have probably launched in less than perfect health. (No one wants to miss a flight and have to go to the end of the line, and most astronauts would lie, cheat, and steal to avoid that fate.) As for the commanders and MSs, they are highly specialized on the primary payload and would be the most difficult to replace on short notice.

## Are astronauts prohibited from doing things that might injure themselves?

The Astronaut Office does have rules to minimize the possibility that a crewmember in training might get injured off the job. These rules say astronauts cannot engage in high-risk activities within 1 year of their launch and medium-risk activities within 8 months of launch. High-risk activities include air racing, snow skiing, and other similar sports. Medium-risk activities are defined as water skiing, softball, and other team sports. NASA takes these rules seriously, as astronaut Robert (Hoot) Gibson found. In 1990, while assigned to a mission, he engaged in some formula-one air racing in his own garage-built midget racer. During one race, another pilot collided with Gibson's aircraft. That pilot was killed and Hoot narrowly escaped death. But he didn't escape the wrath of Dan Brandenstein, Chief of the Astronauts, for violating Astronaut Office policy. As punishment, Hoot was removed from his shuttle mission and grounded from flying the NASA T-38 jets for 1 year. (It obviously was a temporary setback for Hoot. He went on to become the Chief of Astronauts and to command the first *Mir*–shuttle rendezvous mission.)

### Can astronauts fly their training aircraft before a launch?

Yes. There are no restrictions on flying the T-38 jet trainers at any time prior to launch. They are trainers and do help astronauts keep razor-sharp reactions in stressful situations—an important part of preparing for a shuttle flight. Some people, though, have argued that T-38 flying should be restricted in the weeks immediately prior to a launch since the astronauts expose themselves to accidents that could injure or kill them (and jeopardize a mission). But, even on the day prior to a launch, the mission crew can be found cloud-dancing in their jets over the nearby Atlantic Ocean. I suspect there would be an astronaut revolt if any bureaucrat ever tried to ground astronauts from flying the T-38 when they were close to launch. Besides the legitimate training aspects of jet flying, flight is life to most astronauts, and they will fight to prevent anyone from restricting their access to the sky.

### What is crew quarantine?

This is a program designed to reduce the chances of a crew member carrying a cold or flu or similar infectious virus on a mission.

Obviously, if someone became ill in space, their performance could be negatively affected. Also, they could pass the bug to others. To reduce this possibility, the crew quarantine program was instituted. Basically, it is designed to keep the crew as germ-free as possible for the last week prior to their mission. Contact with the quarantined astronauts is limited only to mission-essential personnel and they also have to be free of any symptoms of colds or other sicknesses. If there is any doubt about a person's health, they will wear a face mask when they are in the company of the crew. Also, when the astronauts move from building to building for their training, a security guard will precede them to notify people to stay clear of the astronauts (sort of reminds me of the approach of Biblical lepers).

**Can astronauts stay with their families when they are in quarantine?**

If a quarantined astronaut has children at home who are under the age of 18, the astronaut is prohibited from staying at home and must reside in a motel-like facility at the Johnson Space Center. (Children are considered a greater disease-carrying risk than adults.) The last time an astronaut mother or father will get to say an intimate good-bye to their under-age son or daughter will be 7 days prior to launch. Also, if a spouse has symptoms of illness, he or she will not be able to see their astronaut husband or wife or, if they are allowed a good-bye kiss it will be with a surgical mask on their faces—a new form of safe sex. Astronauts with no at-home children and/or no spouse can continue to reside at home, but most elect to stay with the rest of the crew at the quarantine facility.

There are two crew quarters where the quarantine is enforced: one at the Johnson Space Center and one at the Kennedy Space Center. Three days prior to launch, the crew will leave the Johnson Space Center quarantine facility and travel to Kennedy Space Center and stay at that quarantine facility.

**Do astronauts change their sleep schedules before launch?**

Yes. For example, if the launch time is 3 AM, the prelaunch crew sleep period will be from approximately 2 PM to 10 PM.

The crew quarters bedrooms have no windows, so darkness is assured when sleep is scheduled, and the communal living room has a special, extra-intense lighting system to trick the brain into believing it's day when the awake time is scheduled. Supposedly, this makes it easier to adjust to the mission-required sleep/awake cycle. But some astronauts have saved the taxpayers money on electrical bills by making their circadian shifts with the help of a couple shots of Wild Turkey.

**Are astronauts quarantined after landing from a mission?**

No. Some people confuse the shuttle-era quarantine described above with the *Apollo post landing* quarantine. The *Apollo* quarantine was done after the first several *Apollo* missions to ensure the moon walkers didn't bring back some type of Andromeda strain virus that could wipe out all of humanity. No such threat exists on shuttle flights.

**How many people from all countries have flown in space?**

At the writing of this book, 336 different people have been in space.

**How many Americans have flown in space?**

Two hundred fourteen Americans have been in space.

**How many Russians (and former Soviet Union cosmonauts) have flown in space?**

Eighty-two Russians/Soviets have been in space.

**What other nationalities are represented in the astronaut/cosmonaut ranks?**

Forty individuals from the following nations have flown in space aboard Russian and American spacecraft: Germany (the former

West and East), France, Canada, Japan, Bulgaria, Poland, Hungary, Vietnam, Cuba, Romania, Mongolia, Mexico, Holland, Saudi Arabia, India, Syria, Afghanistan, the United Kingdom, Austria, Belgium, Switzerland, Italy, and Czechoslovakia.

### How many American men have flown in space?

One hundred ninety American men have flown in space.

### How many American women have flown in space?

Twenty-four American women have made space flights.

### How many Russian women have flown in space?

As of November 1995, two Soviet/Russian women have flown into space.

### Who was the youngest astronaut to fly in space?

Sally Ride was 32 years old on her first flight (STS 7, *Challenger*, June 18, 1983), making her the youngest American to ever fly in space. Yuri Gagarin, at age 28, was the first human and the youngest person to ever make a space flight (*Vostok* 1, April 12, 1961).

### Who was the oldest astronaut to fly in space?

American astronaut Vance Brand was 59 years, 6 months, and 29 days old on his last shuttle mission (STS 41B, *Challenger*, February 3, 1984), making him the oldest person to ever fly in space.

### Who was the first grandfather to fly in space?

Admiral Richard Truly, an early shuttle astronaut who later went on to become a NASA administrator, was the first space grandfather.

## How often does an astronaut fly in space?

Assuming there are no major delays in the current launch schedule of seven or eight missions a year, astronauts will average approximately one mission every 2 years. As of STS-74, the shuttle
astronaut flight experience is as follows:

| No. of Astronauts | Have Flown a Total of X Shuttle Missions |
|---|---|
| 3 | 5 |
| 21 | 4 |
| 44 | 3 |
| 68 | 2 |
| 56 | 1 |

## Which American man has flown the most times in space?

John Young is the most experienced American astronaut. Before
retiring from spaceflight, he flew two *Gemini* missions, two *Apollo*
missions (including a lunar landing), and two shuttle missions.

## Which American woman has flown the most times in space?

Shannon Lucid holds this record, having flown five space missions. She was aboard the *Mir* space station for 188 days, and so
she now holds another record as well. She is the American (man
or woman) who has spent the most time in space.

## How many astronauts are on active duty with NASA?

NASA currently employs 88 astronauts, including 70 men, 18
women, 2 African Americans, and 3 Hispanic Americans. They
are divided into 31 pilots and 57 MSs. All the pilots are military
or former military flyers. Of the 57 MSs, 37 are civilian and 20
are active-duty military. Among this astronaut corps are 25
Ph.Ds, (including 10 women), 11 medical doctors (including 3
women), and 1 veterinarian.

**How many astronauts have walked on the moon?**

Twelve. There were six lunar landing missions, each with two men.

**How many women astronauts have walked on the moon?**

None. During the *Apollo* program, the astronaut corps was all male.

**How many African Americans have been chosen as astronauts?**

Seven: 6 men and 1 woman. Guy Bluford was the first African-American man to fly in space, and Mae Jemison was the first African-American woman to fly. Neither was the first black to fly in space. That first was taken by a Cuban, Arnaldo Tamayo-Mendez, who flew with the Russians in 1980.

**How many astronauts have been to other planets?**

None. There are many photos of our solar system's other planets, but those were taken by robot space probes. The only place humans have ever set foot besides the earth is our moon (and it's a *moon* not a planet).

**How many cosmonauts have walked on the moon?**

None. Only 12 human beings have ever walked on the moon, and they were all American.

**What's a Maggot astronaut?**

Several of the astronaut classes have nicknamed themselves. There is no requirement for this. It just happens. My 1978 class of astronauts was the first group of shuttle-era astronauts, and we spontaneously began to call ourselves the *TFNGs,* for *Thirty-five New Guys.* (There's an obscene double entendre in these initials that most military veterans will recognize.) One of the 1984 astronaut selectees—a Navy officer—began referring to his group

of astronaut recruits by the derogatory term that Marine drill instructors use to refer to their new recruits, *maggots*. The nick-
name stuck and now the 1984 astronaut group is the Maggots.

**What's the scariest thing that ever happened to you as an astro-
naut?**

Every launch countdown is a fearful event, but my first launch
attempt holds my scariest memories. The mission was the twelfth
flight in the shuttle series and the first flight of the orbiter *Discovery*. The countdown proceeded to liquid engine start—at T
minus 7 seconds—and the cockpit was filled with a heavy vibration and a loud, growling noise. Along with the rest of the crew,
I watched the final digits flicker off the countdown clock—6. . .
5 . . . 4. Then, there was silence. We had experienced the first
shuttle post-engine-start launch abort. The computers had detected a malfunction in one of the engine valves and had shut off
all of the engines. (This is the reason the liquid-fueled engines
are started early. Unlike the solid-fueled engines, they can be shut
down if there's a problem.) Unfortunately we, as well as the
Launch Control team, had not been adequately trained for such
a scenario, and there was momentary confusion on the radios. It
was this confusion that generated my greatest fear. Astronauts
place their lives in the hands of the NASA team, and any hint
that team is unsure of a situation will fill a cockpit with fear.
That was certainly the situation in *Discovery's* cockpit. To make
matters worse, a small fire broke out at the base of the launch
pad, and Launch Control turned on the fire-suppression system,
spraying the launch pad with water. (Launch Control reported
the fire as small, but when you are strapped to 4 million pounds
of propellant, there's no such thing as a *small* fire.) This was the
most frightened I've ever been as an astronaut.

**Are astronauts more scared to fly since the *Challenger* disaster?**

I flew one mission prior to *Challenger* and two afterwards. My fear
factor never changed among those missions. I was terrified during

the countdown for my first mission, terrified during the countdown for my second mission, and terrified during the countdown for my third mission. I suspect most astronauts would admit their level of fear (whatever it was), like mine, remained unchanged between their pre- and post-*Challenger* flights. That statement may come as a surprise to many people, but you have to remember where many astronauts of the *Challenger* era came from—the military. Tragedy and death are not uncommon events in the aviation services of the military, even in peacetime. Those of us who were Vietnam aviators saw even more of it. Not to sound callous, but *Challenger* was not *shocking* to the military astronauts. We, like everyone else, were surprised by the *cause* of the explosion— a team failure to react to a known problem with the O-ring design—but we weren't shocked that something as complex as a space shuttle had blown up. We always believed, and still do, that flying high performance machines—be they military jets or NASA spaceships—is dangerous. *Challenger* didn't change the fear associated with that belief one way or the other. If anything, there was a little bit less fear after *Challenger* because we knew the team was less likely to make a mistake.

### Is it more scary to fly on the shuttle than it was to fly in combat in Vietnam?

I flew 150 combat missions in Vietnam and three space missions and felt the pre-mission apprehension was much higher for the shuttle missions than for the combat missions. Sitting on the runway at Saigon's airport and anticipating a tactical reconnaissance mission over the Ho Chi Minh trail did not scare me as much as lying in the cockpit of a shuttle and watching the final digits flicker off the countdown clock. I suspect this is the difference between having a sense of control in an aircraft versus the feeling of being a prisoner of the machine when in a shuttle. But this is just a comparison of pre-mission fear. When it comes to the actual events, combat holds the far greater fear potential. Nothing is more sweat drenching, pulse pounding, and adrenaline producing than actual combat—seeing the glowing tracers of anti-aircraft fire coming toward your aircraft and realizing that somebody is actually trying to kill you. That didn't happen on every combat mission. In fact, it rarely happened. But when it did, it was terrifying to a level beyond the terror of riding a shuttle.

### Which astronauts flew the most dangerous space mission in the history of NASA?

At the risk of incurring the wrath of the early astronauts and moon walkers (who were obviously exposed to incredible dangers), it is my opinion that the most dangerous space mission ever flown was STS-1, the first shuttle flight. The reason I say this is because John Young and Bob Crippen launched aboard a never-tested rocket with a marginal escape system. No earlier space travelers—including Al Shepard, John Glenn, Neil Armstrong and even the *Apollo 13* crew—embarked on their missions until their rocket design had been proved in unmanned tests. Even the lunar lander design was tested in Earth orbit (then in lunar orbit) before it was okayed for use by Neil Armstrong and Buzz Aldrin on *Apollo 11*. On the other hand, when Young and Crippen launched on *Columbia's* first mission, the *only* thing that had positively been demonstrated in *actual flight testing* was that a shuttle—when dropped from the back of a Boeing 747 carrier aircraft—would glide to a landing. Everything else about the shuttle system—the solid and liquid engines, the computer hardware and software, the heat tiles, *everything else*—was certified in ground tests, wind tunnel tests, and computer model tests. Young and Crippen were truly facing unknowns that earlier astronauts never had to face. Not only that, they had to face them with a marginally effective launch escape system. While the earlier astronauts always had reasonably good protection in the event of a launch explosion (their capsules could be separated from an exploding rocket and parachuted to safety), Young and Crippen had SR-71 Blackbird ejection seats that provided minimal protection. It was doubted that these seats could have gotten them safely away from a ground explosion (the fireball would probably have been too big) and the shuttle's acceleration quickly took them out of the altitude and speed envelope for the seat to work. So, in my opinion, Young and Crippen were really hanging it out—more so than any other earlier astronaut—when they climbed aboard *Columbia* for the shuttle's first flight.

### What was the safest shuttle mission ever flown by NASA?

STS-26, the first mission after *Challenger*, was, in my opinion, the safest shuttle mission ever launched. The first manned *Apollo* flight (an Earth orbit test of the then-new *Apollo* capsule) could also be considered a safer flight, if any spaceflight can be so labeled. In both cases, the preceding tragedies—the *Challenger* explosion and the *Apollo I* fire—brutally reawakened the NASA team to their vulnerability. I can't help but believe there was a little extra level of team scrutiny to these missions that made them a little safer than others.

### Was it a rush to fly into space?

Strangely, for me, the real rush of spaceflight didn't come during flight but rather in the weeks *prior* to flight. On several occasions before each of my missions (and particularly prior to my first mission), I would bolt awake in the middle of the night with a sudden and overwhelming realization that I was next! That the next rocket to leave the earth would be carrying me! Apparently, that thought was sufficiently powerful to escape even sleep and bring me instantly awake with a wildly thudding heart. A similar rush never came when I was awake, probably because the chaos of the mission training buried it too deeply. Only in sleep—with training pressures swept from immediate consciousness—could the reality of being next to fly make it to the surface of my brain and bring this rush.

### Is there a NASA exercise program?

No. I have seen supermarket tabloid headlines heralding "Lose Weight by Following the Astronaut Exercise Program." It's a fraud. NASA's only health requirement is that astronauts be able to pass their physical exams. Those examinations are very thorough and similar to what an airline pilot would undergo. How an astronaut stays in shape to pass this checkup is entirely up to the individual. Most astronauts exercise regularly by jogging and playing racquet sports, but the only exercise some engage in is walking to the door to pay the pizza delivery person.

## Is there a NASA diet program?

No. NASA doesn't care what you eat as long as you can pass the physical exam. I have seen astronauts eat a vending machine lunch of a Diet Coke and a Twinkie while others have had beer-and-peanut suppers.

## Are astronauts superstitious?

I do not have any superstitions and have never heard of or witnessed any other astronauts' superstitions. However, that's not to say the space program doesn't have superstitious people. I recall one example of superstitious behavior during a training exercise for my second mission. The flight, STS-27, was to be the second post-*Challenger* mission. I noticed that above the entry hatch of a training simulator was a colorful display of stick-on decals of the mission patches for all prior missions. Seeing that our patch was missing from the display, I added our decal. Several days later when our crew was again in training at this simulator, I noticed our patch had been scraped off. When I asked a nearby NASA employee for an explanation, I was mildly shocked to hear it was rooted in superstition. With one prior exception, the training personnel for this simulator had always added a decal to the display *after* the mission had returned safely, never *before* it was launched. The only exception was *Challenger*, when they had put it on before launch. When they had discovered I had added the STS-27 decal prior to our launch, they had removed it, obviously fearing another *Challenger*. After we landed safely, they added it to the display.

### Are astronauts religious?

The astronauts are a cross-section of America and reflect that cross-section in their religious beliefs (and absence of belief). There are people I would classify as fundamentalist Christians. There are Jewish and Catholic astronauts. At least one astronaut was Buddhist. But there are also agnostic and atheist astronauts.

### Does space flight spiritually change an astronaut?

I suspect those individuals who have a fundamental belief in some higher power have those beliefs reinforced by the incredible majesty and beauty of Earth as seen from space. I personally had my religious beliefs reinforced by my flights into orbit. However, I suspect there have been astronauts who have not been similarly moved.

### What's the astronaut divorce rate?

As of January 1996, approximately 10% of all astronauts selected since 1978 (the shuttle-era astronauts) have divorced while at NASA. While this seems like a very low number compared with the 50% breakup rate that is normally quoted for all marriages in America, my research was very unscientific and cannot be compared with this all-inclusive 50% figure. I merely asked a NASA secretary (they know *everything*) to review the astronaut class photos and count the number of individuals known to have divorced *while still with NASA*. In other words, I did not include any divorces that occurred before an astronaut arrived at NASA or after an astronaut left NASA. Also, I made no investigation of whether the divorces were early in an astronaut's career (implying the marriage was terminal before the couple arrived at NASA). I suspect if you compared astronaut divorce rates with a similar group of non-astronaut couples (well educated, 30 to 50 years old), you would find the numbers very similar.

## Can astronauts buy life insurance?

Yes. There are insurance companies that cater to aviation crew members and that have no clauses to limit a payout in the event of death during a space mission. Also, NASA offers a policy for sale in which the insurance underwriter covers shuttle accidents. Before my first flight in 1984, I wrote a letter to my insurers specifying that I would be flying a space shuttle and asking them to certify I would be covered in the event I was killed. They did and I attached the replies to my will. I didn't want my wife getting any small-print surprises.

## Can pregnant women fly in space?

No. The effects of launch g-forces and weightlessness on a pregnant woman are unknown and could endanger the health of fetus and mother. Also, the effects of space radiation on a days' old fetus are thought to be severe. NASA doctors test women for pregnancy about 1 week prior to flight and will ground them if they are pregnant. This has never occurred.

## Can married astronauts be on the same shuttle crew?

NASA will not assign married astronauts to the same mission. There are two reasons for such a rule. First, if the married couple have children, putting mother and father together on the same flight exposes them to simultaneous death and their children to becoming instant orphans. The second reason is that intimate personal relationships, as exist between husband and wife, can jeopardize sound judgment and decision making. For example, no husband or wife likes to get *ordered* by their partner, yet that might be required on a shuttle mission. While it's possible that a spouse would be able to respond in a professional manner to such orders, there is certainly an emotional complexity to the situation that doesn't exist when crewmembers aren't married. Simply put, flying a married couple on a shuttle adds to the complication of an already complicated business.

As I said above, NASA will not and has not ever *assigned* married astronauts to the same mission, but a married couple has flown in space. The astronauts in question weren't married when they were selected for the flight, but married during their mission training. When this occurred, NASA intended to remove one of them from the flight but after hearing the couple's objections and since they were childless, NASA made a one-time exception and allowed them to fly together.

### How many married astronaut couples are there?

In the history of the shuttle program there have been six married astronaut couples. As of January, 1996, there are four couples still serving as active astronauts.

### Where do astronauts live?

Many people think astronauts live in Florida since the shuttles are launched there. This is not true. All astronauts must live in Houston, Texas because 99% of astronaut training occurs at the Johnson Space Center.

### Are astronauts tested for claustrophobia?

Yes. During the astronaut selection process, each candidate is required to get into a rescue sphere for about 20 minutes. (See Chapter 3 for a discussion on the rescue sphere.) Few things could be more claustrophobic than curling into a fetal position and being zippered into a dark, windowless ball and left alone for an indeterminate time. (Prior to the test, candidates have to remove their watches so they have no measure of time.) I don't know of any candidates who flunked this test, which implies claustrophobics don't apply to be astronauts.

I think the most claustrophobic activity an astronaut regularly experiences are the underwater spacewalk training sessions. Just *dressing* in the spacesuit is a claustrophobic's nightmare. A deep breath isn't even possible without feeling the constriction of the torso. Now, imagine being lowered underwater. If you have any claustrophobia, you'll know it when the water starts coming up to eye level.

## Do astronauts see psychiatrists?

During the astronaut selection process, two psychiatrists interview each candidate. Some of the questions they asked are, "If you died, what epitaph would be etched on your tomb?" (If you said "Good Riddance" it probably would cast a shadow on your selection potential) and, "After death, if you could come back as an animal what type of animal would you want to be?" (Almost everybody said an eagle or a falcon. A wish to be reincarnated as an aardvark probably wouldn't denote an astronaut disposition.) They also gave us oral tests, like having us count backwards from 100 by 7s as fast as possible. (Try it sometime. It isn't easy.) I have no idea how a psychiatrist determines who is fit and who isn't through such questions and tests and, frankly, I doubt anybody was eliminated in these interviews because of their performance. After being selected as an astronaut, psychiatric help remains available to anybody who wants it. No figures are maintained on how many astronauts have availed themselves of this help.

## Do astronauts always get along with each other?

No. At any one time, the Astronaut Office has about 90 people in it. When it comes to interpersonal relationships, it's no different than any other similar-sized office—some people get along, others do not. On very rare occasions, this friction surfaces in an explosion of conflicting personalities, for example, on the eve of one mission a commander was heard screaming at the top of his lungs in the crew quarters at one of his crewmembers.

## Do astronauts party together a lot?

In the first few years the bonds of an astronaut class are very tight and there are frequent parties and other get-togethers. As the years pass, however, human nature begins to erode these bonds.

People get to know each other too well, and sometimes don't like what they see. Also, flight assignments can have a corrosive effect on astronaut relationships—Why did she get that space walk and I didn't? Why is he flying ahead of me? Astronauts aren't robots. We are afflicted with the same trappings of humanity—insecurities and jealousies—as everybody else. So, over the years, class parties become less and less frequent until they cease altogether. In the end, astronauts socialize, as people everywhere do, with the subset of their coworkers they consider to be close friends.

### Is there an astronaut hang-out away from NASA?

Probably the most frequented astronaut hang-out is the Outpost Tavern, near the Johnson Space Center. Gene Ross, the owner, is a great NASA friend and has hosted countless shuttle-era astronaut gatherings. If you ever go for a visit, though, don't look for a grand (or even *modern*) edifice. The Outpost looks like something from a Wild-West movie set, complete with saloon doors shaped as curvaceous cowgirls and holes in the unpaved parking lot that can swallow whole automobiles. For all the astronaut emotions that its smoky interior has witnessed (STS-1 was celebrated and *Challenger* was mourned there), I think it should be a registered National Historic Landmark.

### Is there an astronaut band?

Yes. Named *Max-q*, an aeronautical term meaning *the point of maximum aerodynamic pressure*, it is composed of a handful of musically inclined astronauts and frequently plays at astronaut parties. If you are applying to be an astronaut and have a talent for music, you might just want to put it on your résumé. Who knows, it might end up being the tie breaker that gets you in.

### Do the astronauts spend the night before launch with their families?

No. When astronauts arrive at the Kennedy Space Center (usually 3 days prior to launch), they stay in the Astronaut Crew

Quarters. This is a motel-like facility that has been built into a portion of the third floor of a large office/satellite checkout building. It has bedrooms, meeting rooms, a kitchen, and a small gymnasium. The purpose of the crew quarters is to continue the health quarantine that began at the Johnson Space Center at L-7 days (7 days prior to launch) and give the crew members a retreat for relaxation and any final checklist studying they may want to do. There is cable TV and a small library of video movies to watch. I always got a kick out of the fact that sometimes astronauts watch cable TV from satellites *they* put in orbit.

Families are not allowed to stay in the crew quarters. Spouses, however, are allowed to visit the astronauts and share some meals with them at the discretion of the crew commander. Usually the last spouse–astronaut contact occurs at the Astronaut Beach House about 24 hours prior to launch. After that, spouses are usually not permitted to come to the crew quarters.

### What's the Astronaut Beach House?

Prior to the beginning of the space race in the late 1950s, there were private homes on what is now the Kennedy Space Center. In NASA's moon-race expansion of their launch facilities, the US government bought out the homeowners and razed their properties. Only one small house, situated on the beach, was spared destruction, and it was converted into a small conference retreat. It is known as the *Astronaut Beach House* because mission crews use it as a rendezvous point to say good-bye to spouses in the days prior to launch. (Those wishing to read more about the unique and dramatic place this building occupies in space history should read my article, "The Beach House," in the June/July 1992 *Air & Space Smithsonian Magazine*.)

### Do astronauts really eat a breakfast of steak and eggs on launch morning?

Maybe the original astronauts ate big breakfasts like steak and eggs, but I've never seen a space shuttle astronaut eat such a

meal on launch morning. In fact, I've never seen a shuttle astronaut eat a big breakfast of *anything*. Rookie astronauts, who
are unsure if they are susceptible to space sickness, will usually
eat light (if they eat anything at all), out of fear they will be
vomiting in a couple hours. Experienced astronauts who know
they will be sick obviously are selective about what they eat.

**Author and the rest of the STS-36 crew at the
launch morning breakfast.**

And most astronauts
have reduced appetites
because of the sheer ex-
citement and fear of
being close to launch. I
never ate more than a
piece of toast. Really,
the breakfast is more of a
photo opportunity than a
meal. NASA photogra-
phers briefly film the
crew sitting at the table.

Personally, I always felt uncomfortable during this filming. In the
back of my mind, I wondered if this was to be one of those last
shots of the crew before the accident that killed them.

Besides avoiding food on launch morning, most experienced
astronauts will limit their consumption of liquids. They know in
a couple hours their bladders will be overfull. What's the big deal
about that, you ask? Won't everybody be wearing a urine collection device? Yes. But it's difficult (particularly for men) to urinate lying on their backs, even when their bladders seem ready
to burst. Also, it's uncomfortable to lie in a wet diaper, so most
people are careful about what they drink before launch. I've
never seen anybody drink a cup of coffee on launch morning.
That would be bladder suicide. In fact, many astronauts have a
dehydration ritual they follow—like jogging or exercising
(sweating is the objective) the night before a mission and not
drinking much before bed. I once saw an astronaut on the eve
of a launch sitting in a hot whirlpool drinking a beer. He had
already jogged and was sweating but wanted the diuretic effect
of the alcohol and heat to make him dryer yet.

## When are you officially an astronaut?

The word *astronaut* translates (from Greek) to mean "star voyager." Fortunately, however, the official requirement to be an astronaut doesn't require travel to another star (the closest one is about 30 *trillion* miles away). All you have to do is get to 50 miles altitude. That's the official definition of an astronaut—somebody who has flown 50 miles above the earth.

Most people think the only astronauts have been people who have flown NASA rockets, but this is not true. A handful of people earned astronaut wings in the 1960s and 1970s by flying the X-15 rocket plane above 50 miles altitude. In fact, Neil Armstrong and Joe Engle were X-15 astronauts *before* they ever flew NASA rockets.

I find this official definition just a little crazy. It would mean officially the three rookie crew members of *Challenger* are not astronauts. Though they died aboard a NASA rocket, Mike Smith, Christa McAuliffe, and Greg Jarvis never got any higher than about 10 miles altitude (where the explosion occurred). Who's going to say the *Challenger* crewmembers are not official astronauts? I wouldn't. Also, it would be possible in some shuttle launch emergencies for a crew to not reach 50 miles high, even though they might almost die bringing a crippled shuttle back to Florida. Would you like to tell them they didn't earn their astronaut wings on an aborted mission that almost killed them?

I think we agree. Altitude is a silly way of defining an astronaut. When you're flying a rocket, you earn your astronaut wings at lift-off, regardless of how high you ultimately get.

## When does NASA designate someone an astronaut?

When a new group of astronauts first comes to NASA, they really aren't called astronauts. They are designated astronaut candidates. (For this reason, brand new astronauts are called *ascans*—pronounced "ass cans.") After about a 1-year evaluation period, NASA changes their title to *astronaut*. At this point the agency feels reasonably certain they will fit in and can bear the prestigious mantle of astronaut without embarrassing the agency. (In other words, it's easier to fire somebody who is an astronaut candidate than it is to bounce an astronaut.) To the astronauts, though, this titling is just a game of names. Regardless of what you are called and when you are called it, we believe you're not really an astronaut until lift-off in a spaceship.

**What's an astronaut pin?**

It's a small, gold lapel pin astronauts receive after their first flight into space. There is a silver astronaut pin given to astronauts after their year of candidacy is over, but few astronauts wear these. As far as the astronaut corps is concerned, the title *astronaut* means nothing until you've actually been launched. So wearing a silver astronaut pin would be like an airline passenger wearing Delta wings. They may look nice, but they're meaningless.

The pin design consists of a three-rayed shooting star passing through an ellipse. The *Mercury* astronauts came up with the design, and they did a great job. It's a simple, yet elegant design. All astronauts should give thanks that there was some artistic talent in the *Mercury* group. The other examples of astronaut art—our mission patches—are sometimes hideous.

The astronaut office has the gold pins flown aboard shuttles before they are awarded, so they are definitely unique. No other lapel pins have been in space. The pins are usually given at a landing party (a celebration of a successful shuttle mission).

**How much do astronauts get paid?**

If you asked most astronauts, they probably couldn't tell you exactly how much they are paid. When you're living a dream, you don't think too much about money.

But let's answer the question. What are astronauts paid? First of all, you must understand that being an astronaut is a government job, so astronauts are paid by the government. How much they are paid depends upon a couple things: years of government service and any specialized training. An old-head, Ph.D. astronaut with 20 years of service will be paid about $91,000 per year (as of 1995). A new astronaut with minimum experience in his field will be hired as a GS-13, and his pay will be about $56,000 a year (as of 1995). But that will be exactly the same as a GS-13 working in the Department of Agriculture or in the Department of Commerce or anywhere else in the government. Astronauts are not paid a bonus just because they are astronauts.

Anybody hired from the military services to be an astronaut will continue to be paid by their service depending on their rank. Their pay will not change merely because they have become astronauts.

Astronauts are paid extra when they fly in space, just as anybody who travels on government business is paid extra for meals, hotels, taxis, laundry, and so forth. When they return from orbit, astronauts fill out a form saying where they've been (in space), what taxi they took (a government shuttle), where they stayed (in a government cockpit), and what they ate (government space food). Generally, they get an extra $20 or $30 dollars in per diem for the mission. Astronauts laugh about this and wish NASA would just pay them a penny for every mile they've traveled (400,000 miles per day at 1 cent per mile equals $4,000 per day).

### How many American astronauts have died from all causes?

As of November 1995, 24 American astronauts are deceased.

### How many American astronauts have died in accidents?

Thirteen astronauts have died in line-of-duty accidents, including three aboard the Apollo I capsule, five aboard the shuttle Challenger, four in T-38 aircraft crashes, and one shuttle astronaut who died in a commercial plane crash while traveling for NASA to give a speech. Additionally, two non-NASA employees lost their lives on Challenger: Greg Jarvis and Christa McAuliffe.

Five other astronauts or retired astronauts have died in non-NASA-related accidents: one in a car crash, three in private aircraft crashes, and one from altitude sickness while climbing Mt. Everest.

**How many astronauts have died of natural causes?**

One astronaut died of cancer while still on active duty with NASA. Six others have died of natural causes after retiring from NASA: two from cancer, one from complications of pancreatitis, and three from heart attacks.

**Did the Russian strand a cosmonaut in space?**

In the months surrounding the Soviet Union's dissolution there were a lot of press reports about a Russian cosmonaut being stranded in space aboard the Mir space station. This never happened. The Russians always leave a Soyuz return capsule docked to their space station in case the cosmonauts ever had to abandon the Mir in an emergency. The cosmonaut in question merely had his visit involuntarily extended to accommodate some scheduling changes with other Mir visitors. The only newsworthy item in this drama was that during his stay in space, the cosmonaut's country (the Soviet Union) ceased to exit.

**Are astronauts really like the ones in the movie The Right Stuff?**

The vast majority of astronauts are goal-oriented, type-A individuals and for that reason they can be very intense and opinionated, both of which are personality traits captured in the characters of The Right Stuff movie. However, in my opinion, the movie also paints astronauts as being excessively self-absorbed and arrogant. In my career, I observed some of these, but the majority of shuttle astronauts are true team workers willing to listen to others and participate in a consensus decision.

**How realistic was the Apollo 13 movie?**

Very. The movie closely followed the actual events of the ill-fated mission. It also did a great job of capturing the *real* heroes of the mission—the Mission Control team. Most space movies always focus on the astronauts. Apollo 13 was realistically refreshing in that it highlighted a Mission Control team that saved the astronauts through incredible acts of teamwork and ingenuity.

Another aspect of the movie that was dead on target was the stress on astronaut families. Over the years there has been a river of tears shed by families watching loved ones leave the earth in spaceships. This drama is mostly invisible because it occurs in private, family moments, and it's not Right Stuffish to share it with the public. But it's there, and *Apollo 13* captured it.

Film Director Ron Howard also went to great efforts to make the weightless scenes as realistic as possible. The studio paid for the NASA Vomit Comet to fly the actors, and they were filmed in 20 to 30 second weightless bits. Tom Hanks and the others endured 596 weightless parabolas over 17 flights to get enough film to use in the story.

### Do retired astronauts have reunions?

Yes. Every 2 years (it used to be every year) a reunion dinner is held in Houston, Texas. I would estimate only about one quarter of retired astronauts attend these functions. Those that do have a good time trading stories (and teasing each other about ballooning waistlines and receding hairlines).

### Is there an astronaut society?

Yes. The Association of Space Explorers (ASE), is an international organization with perhaps the most unique entry qualification of any club in the world. To join, you must have flown into space. The ASE's mission is to promote space exploration to the world's citizens, and each year its astronaut and cosmonaut members hold public meetings in various places around the world to further this mission.

### Are astronauts famous?

Not any more. In the earliest days of the space race, astronauts were so famous, some people wouldn't cash their checks because they wanted to keep their signature. (That's the type of fame I would like!) The reality now is no shuttle astronaut is famous in the sense they have name or appearance recognition. (And everybody cashes their checks.) Actually, the only astronaut name recognition I have found in my dealings with the public are of Neil Armstrong and Sally Ride and, even then, many people won't know exactly what they are famous for. (Neil was the first man on the moon. Sally was the first American woman in space.)

### Do people write to astronauts asking for their autographs?

Yes. On average, an active astronaut will get between 5 and 20 autograph requests per week from people of all sorts: young children, collectors, foreign nationals, and so on. NASA has no requirement that astronauts honor these requests, but most do. By law, no active astronaut can sell an autograph.

### Do astronauts really say "A-OK"?

The term *A-OK*—meaning all systems are okay—is NASA slang from the early space program. No shuttle astronaut uses it. In fact, its usage today would mark you as some type of space dweeb. Shuttle astronauts use the terms Go, Green, or Nominal, as in, Everything's go, or, Everything's in the green, or Everything's nominal.

### What made you retire from the astronaut program?

This is a frequently asked question and difficult to answer. Most of my life I dreamed of being an astronaut and yet I voluntarily walked away from it after my third shuttle flight. Why? The reasons are manifold. First, there was the *extreme* frustration of never really knowing what management had in store for you. Astronauts were largely kept in the dark about their futures. Would there be a fourth mission? When might it occur? Answers to such simple, life-planning questions were treated by some managers in the astronaut chain of command as if they were nuclear secrets. Second, even if I had been given absolute assurances there would be a fourth mission assignment, there was the fear that it might never happen. On the eve of a fourth flight—after a 2 year investment of my life—I could fail a physical exam or the shuttle program could suffer another major delay. Third, there was fear I could die. As much as astronauts would like to say they don't experience fear, it would be a lie. I had already tempted fate with three missions; did I want to take the risk again? Then, there was the stress my career was exerting on my family. Did I want to put them through another countdown? All of these factors influenced me to decide three times in space was enough.

I should include this observation on astronaut retirement. In a way, it's the most fearful thing an astronaut will ever face, more fearful than any mission. It's impossible to put aside this thought—that retirement to any second career is a step down on the ladder of life. No amount of fame or money gained in any post-astronaut career will ever buy another ticket into space, and for that reason, every retiring astronaut knows, in a strange way, they will forever be paupers. But, such is life. Every astronaut eventually faces that other great unknown called retirement.

### Do astronauts get to keep their spacesuits when they retire?

We are not allowed to keep the orange suits we wear for launch and reentry. These are pressure suits that cost hundreds of thousands of dollars and are used over and over for various crew members. The real spacesuits (i.e., the white suits that spacewalkers wear) cost millions of dollars and weigh about 275 pounds. NASA isn't about to let these go, either. The only spacesuit we get to keep isn't really a spacesuit. It's the blue, patch-emblazoned coveralls that we wear when flying the NASA T-38 training jets. These are tailored to each individual, so NASA has no use for them.

### Do you miss being an astronaut?

Yes. There's always a twinge of loss whenever I see TV shots of people driving out to the launch pad or floating around the cockpit. I wish there were some way to snap my fingers and be in the picture. But this doesn't mean I regret my decision to retire. My post-astronaut life has been very rewarding and emotionally enriching. It's just human nature to be nostalgic about something as powerful as the experience of spaceflight.

### What do retired astronauts do?

Many astronauts remain with NASA in managerial positions. Others take jobs with NASA contractors, and some of these have risen to chief executive officer status. Some start their own consulting companies. A few have gone into politics or returned to universities to teach. One—Al Bean—has become an acclaimed painter. Others have written novels and nonfiction books. A very small minority have somehow suppressed those obsessive-compulsive urges and have completely retired to play golf, fish, and travel. As for myself, I have embarked on a completely different career as an author and professional speaker.

### What did your family think of you being an astronaut?

My wife and children were frightened by the risks involved, but they understood spaceflight was a dream come true for me. They were very supportive but are happy I have retired from it. My wife answers this question by saying, "I wouldn't take a million dollars for the memories, but I wouldn't pay a nickel to do it over."

### Did any of your children want to be astronauts?

I have three children—a son and two daughters. None wanted to pursue an astronaut career. Each one has different talents and different dreams.

So far there have been no offspring of astronauts who have gone on to be astronauts themselves. But I expect that will change as more and more astronaut children grow into adulthood and pursue their own space destinies.

**Did you walk on the moon?**

No. The shuttle can't fly to the moon. Only the *Saturn V* rocket was capable of carrying men to the moon, and it's no longer manufactured.

**Would you liked to have walked on the moon?**

Absolutely! But the moonwalkers are the only early astronauts I envy. With the exception of the *Skylab* astronauts, all the preshuttle astronauts flew in small capsules that had limited visibility. If your entire space experience is going to be in Earth orbit flight (and not on the moon), I would rather be in a vehicle like the shuttle, which has many windows so you can enjoy the sights.

**Are you a Trekkie?**

No. I've never really gotten into the *Star Trek* movies and TV series. My favorite science fiction movies are ones from the 1950s—*War of the Worlds, Forbidden Planet, When Worlds Collide, Invasion of the Body Snatchers*, and so forth. In other words, I like more frightening plots than *Star Trek* delivers. The last good, contemporary sci-fi movie I've seen was the first *Alien* movie. Having baby aliens explode out of an astronaut's chest is more my type of sci-fi entertainment. (One astronaut mimicked this scene while eating lunch in the NASA cafeteria with a group of fellow astronauts. He had an inflatable bag under his shirt that swelled his chest and made it appear something was about to jump out.)

**Do astronauts play jokes on each other?**

Constantly. For example, Judy Resnik once opened her purse to find a live grass snake inside. She got even with the culprit by putting a stinking dead frog in his desk drawer—a big Texas toad that had been flattened by cars until it was as thin as a piece of paper and as wide as a manhole cover. Other astronauts have come to work and found their complete office, including desk, cabinets, chairs and wall hangings, inside the elevator. Still others have returned from a training session and found that their car has been manhandled into the lobby of a building. Astronauts are constantly trying to out do each other with pranks. In a way, such sophomoric activity is part of the culture of being an astronaut. It takes the edge off a job that can be very stressful.

# CHAPTER 9

---

# The Future

## Will NASA ever fly another teacher in space?

The tragedy of *Challenger* was compounded by the fact that one casualty—Christa McAuliffe—was a non-mission-essential passenger, the first teacher in space. After the disaster, NASA realized it was morally indefensible to place people in harm's way for secondary objectives, however noble those objectives might be (Christa's was to excite American children about space and science education), so they indefinitely suspended the teacher-in-space program. Someday, they may decide the shuttle's reliability is sufficiently demonstrated to again send a teacher into space, but such a mission has not yet been scheduled. If it is, Barbara Morgan (Christa's backup) would be the teacher to fly.

More than just the teacher-in-space program was suspended by the *Challenger* tragedy. NASA also had plans to fly journalists, musicians, and artists to better capture the experience of spaceflight for the common man and woman (astronauts struggle to effectively communicate the grandeur of the adventure). Those programs were also indefinitely postponed after *Challenger*.

## Will NASA ever fly kids in space?

When I was a child, at the dawn of the space race, one of the much-discussed news items was the limited lifting power of the early American rockets. It was grist for my childhood dream-mill that NASA would have to use kids as astronauts since adults would be too heavy. Of course, in my dream, I would be one of those kids. Obviously, NASA didn't select children. Children and the rest of the public will have access to space only when technology reaches the point at which space travel is as safe and routine as air travel. I doubt this will happen in my lifetime.

## What is NASA's budget?

Most people believe NASA consumes a significant portion of their tax dollars. Not so. Since 1977 it has averaged less than 1% of the federal budget. For fiscal year 1996 NASA has been authorized approximately $14.2 billion in spending. For comparison, the Health and Human Services budget for the same year is over $600 billion, and the Department of Defense budget exceeds $250 billion. In other words, Health and Human Services consumes the entire NASA budget in about 6 federal workdays. People who argue that the money spent flying into space could be used to solve all our social ills in America should consider this fact: Does anybody really believe an extra 6 days of work by the social welfare agencies would solve our problems?

## Why do we need to fly people into space?

There are two basic reasons why NASA sends people into space: science and money. Many sciences are furthered by observations in space, particularly astronomy, oceanography, meteorology, and geology. While many robotic spacecraft do a marvelous job without a space-based human around, there are limits to their powers of observation. They can't readily adjust to fleeting conditions; for example, how do you tell a robot to look for a plankton bloom on the sea surface or watch for a unique flash of vertical lightning? This is why oceanographers, meteorologists and geologists *crave* astronaut observations and photography, even though they have access to robot-taken photos. Also, there are severe limits to a robot's power of self-repair. Imagine telling the Hubble Space Telescope to repair itself or commanding a stranded communication satellite to heal itself. (NASA has rescued several stranded unmanned satellites and, in the process, has saved the owners hundreds of millions of dollars.)

On the money side, NASA sends people into space with the objective of commercializing it—to demonstrate to American businesses that products can be made in space that will have a terrestrial market. If companies find they can make a profit from a space product or service, they will rush to invest in the appropriate satellites or space factories and the American economy and society will ultimately be enriched by that investment (communication satellites and cable TV are a good example of the private sector exploitation of space). Before this can happen, however, NASA has to explore various product and service possibilities. Private companies can't do it because the costs are too high. Basically, we taxpayers are providing space seed money (through NASA) to investigate the breakthrough possibilities of space-based processes. Specifically, NASA wants to investigate the commercialization aspects of the two unique aspects of orbit flight: weightlessness and an exceptionally pure vacuum. Since various substances exhibit totally different behaviors in such environments than they do in Earth laboratories, it may be possible to make things in space that cannot be made on Earth and that will revolutionize Earth markets (e.g., new drugs, alloys, and computer chips). But a human presence is needed to conduct these experiments. It cannot be done by robotics any more than Earth-based invention can be done solely by robots. We need a laboratory in space (a space station) where human researchers can observe and experiment, make changes, and experiment some more. Will many of these experiments lead to useless dead ends? Absolutely. Will some hit pay dirt of unimaginable value? Probably. That's the nature of research and development. There are no guarantees. But the economic consequences of not trying and letting other countries find those potential gold mines are too severe to accept.

### What are NASA's plans for space exploration?

NASA's first priority is to build an international space station. Next, it intends to return astronauts to the moon and ultimately build a permanent research facility there. Finally, NASA hopes to send astronauts to Mars. All of these projects are technically feasible but all of them—and particularly those that will send human beings back to the moon and on to Mars—will be very expensive. It's unlikely that any of these projects will be adequately funded for many decades. However, this doesn't mean that there won't be some spectacular space research conducted in the next decade by robot spacecraft. Even now, the Galileo spacecraft is exploring the planet Jupiter and its moons from Jovian orbit, and in the fall of 1996, NASA intends to send two spacecraft, *Mars Global Surveyor* and *Mars Pathfinder*, to Mars. *Surveyor* will observe the planet from orbit, while *Pathfinder* will land on the surface and release a small, wheeled rover. The rover will be able to travel about 200 feet per day and is to have a minimum life of 30 days. And these probes of Mars are just the beginning. For the foreseeable future, NASA intends to send at least one robot spacecraft to Mars every time there is a favorable launch window—that is, approximately every 26 months. Eventually NASA hopes to land a spacecraft on Mars that will return a soil sample to earth. There are also long-term plans to send robots to study and return to earth soil samples of comets, asteroids, and the moons of Jupiter and Saturn. NASA will also continue its efforts to better understand the universe through observations by the Hubble telescope and other deep-space-viewing robot spacecraft. A very distant NASA dream is to build a telescope that will be powerful enough to photograph the planets that revolve around other stars.

### When will the space station be built?

The first piece of the international space station is currently scheduled to be launched by a Russian rocket in November 1997. After approximately 14 other Russian launches and 27 shuttle launches, the space station is to be completed by June 2002, at which time it will be permanently occupied by six persons from various countries. NASA is currently flying the shuttle to the Russian space station *Mir* to refine the procedures for building this entirely new international space station.

*The Russian Mir as seen from an approaching space shuttle.*

### Why doesn't NASA's space station design look like a spinning wheel?

Remember all those science fiction movies where the space stations were spinning wheels? The centrifugal force of the spin produced an artificial gravity that allowed astronauts to walk around. These designs were the product of ignorance. When they were first being hypothesized—well before the first manned space missions—nobody had a clue how weightlessness would affect astronauts. There were worries that they could get stranded in the middle of the spacecraft or suffocate in a hovering cloud of their own exhalation, or not be able to eat, drink, sleep, or use a toilet. Also, there were some suggestions that astronauts would go insane in the free fall of weightlessness. (What could be scarier than an infinite fall?) Artificial gravity seemed like a good idea at the time, thus the spinning wheel space stations.

Now, however, artificial gravity is the last thing scientists want on a space station. We now know it's not required for the astronauts, and the whole purpose of such an orbiting laboratory is to see if it's possible to make things in space that can't be made on Earth, so absolute weightlessness is needed. Any artificial gravity would ruin many experiments. In fact, on some shuttle experiments, the crew is asked to curtail their motion and avoid pushing on the interior cockpit surfaces to minimize even minute disturbances to weightlessness. That's not to say someday artificial gravity won't be desired. For example, on long trips through space, it may be necessary for crew health—to keep bones and muscles in good condition or perhaps to facilitate fetal development or childbirth. For now, though, we want a space station to exploit the uniqueness of weightlessness.

### When will NASA send astronauts back to the moon and to Mars?

NASA has no specific dates set for a return to the moon or for a Martian expedition. They want to complete the space station before trying to get funding for these projects. Personally, I doubt another human lunar landing or a first-time Mars visit will occur in my lifetime. The enormous expense of human flight beyond Earth orbit and our smothering deficit will prevent such grandiose undertakings. However, NASA will continue to use robotic space probes, including landers and soil-return vehicles, to explore the moon, planets, asteroids, and comets.

## How long would it take astronauts to fly to Mars?

With current chemical rocket propulsion technology, it takes about 9 months to reach Mars. Someday, shorter travel times might be possible with nuclear rockets.

## What will replace the space shuttle?

The next generation of manned launch vehicle will most likely be a *single-stage to orbit* (SSTO), vehicle. The shuttle is a marvelous, reusable spacecraft, but its operation is complicated (and made significantly more expensive) by the need to lift it with jettisonable booster rockets. Also, every flight requires the expenditure of the external fuel tank (it burns up in the atmosphere). This process of successively getting rid of weight is known as staging and is necessary because the materials from which current rockets are constructed are so heavy. But staging is very expensive. For example, even though the shuttle's solid rocket boosters (SRBs) are reusable, they must be picked up by boats, towed to shore, disassembled, sent back to their Utah factory to be refurbished, retransported to the Kennedy Space Center, and then stacked on the launch platform. All of this costs money. Wouldn't it be wonderful if a craft could be built that would take off, fly into orbit, and return without ever requiring staging? Upon landing, it would take only a few days to load another payload, fill it with fuel, and send it back into space. This is the SSTO dream, and right now it's just that—a dream.

On July 2, 1996, the first step toward making SSTO a reality was taken when NASA announced that the Lockheed Martin Corporation had won the contract to develop the first one. This winged vehicle, which will be named *VentureStar*, will resemble the shuttle in that it will be launched into orbit vertically and return to earth as a glider. *VentureStar* could make its first flight as early as 2003, and the plan is to have it replace the shuttle eventually. However, before that can happen some significant advances will have to be made in rocket engine technology and lightweight material design. If everything goes as dreamed, *VentureStar* will reduce the cost of putting payloads into low earth orbit from the current $5,000 to $10,000 per pound to about $1,000 per pound.


## Will the Russians ever launch a space shuttle?

They already have. Called *Buran*, it flew a brief, unmanned orbit test mission. The Russian now admit, in an expensive act of Cold War paranoia, they built it simply because America had one. Now, with the Cold War over and lacking the resources to keep it flying, the Russian shuttle is essentially grounded. That country continues to use their well-proven, multistage, expendable rockets, similar to our *Titan* rockets, to launch their manned missions.


## Will there ever be spacecraft that can travel beyond our solar system?

To fully appreciate the question of travel beyond our own star system, consider these facts. The speed of light is approximately 186,000 miles per second. The shuttle flies at just under 5 miles per second. At light speed, the closest star—Alpha Centari—is 4.3 years away. At shuttle speeds, that distance would require approximately 158,000 *years* to traverse! And that's the *closest* star. Clearly, any "Star Trek"–type intergalactic exploration in a single human generation will require speeds of many times the speed of light. Is that possible?

No, or so said Albert Einstein. By his theory of relativity (which I don't profess to understand), the energy needed to propel an object to the speed of light is infinite. In other words, it would take more energy than exists in the entire universe to propel a Federation starship. Trekkies, no doubt, see a future in which advances in physics will make Einstein's theory as obsolete as the Wright brothers made "man will never fly" pronouncements obsolete. But, since Einstein's ideas have been repeatedly supported by observations of many different types— from subatomic behavior to the behaviors of galaxies—I suspect the woolly haired gentleman is correct. Humans will never fly at the speed of light—and Star Trek–type of travel will forever be the stuff of science fiction.

Having said this, however, I do believe interstellar travel might someday be possible at more conventional speeds of mere millions of miles per hour. Perhaps in a future Earth generation that's facing annihilation from a cometary impact, an incredibly massive space city with a population of hundreds of thousands could be assembled and sent on a one-way mission from earth. Perhaps its environment and food stores could be made completely renewable. Perhaps it could be resupplied with energy by capturing atoms of errant hydrogen streaming through space. Or maybe it could be built on a captured asteroid that is mined for energy. Perhaps this last vestige of humanity would someday stumble across a habitable planet or one that could be fully terraformed. It takes a significant suspension of disbelief to imagine such a trip could ever be possible, but I can believe this scenario more easily than I can believe a rewrite of Einsteinium relativity.

# Index